T0249832

Applications of Quantum and Classical Connections in Modeling Atomic, Molecular and Electrodynamic Systems

Applications of Quantum and Classical Connections in Modeling Atomic, Molecular and Electrodynamic Systems

Alexandru Popa

AMSTERDAM • BOSTON • HEIDELBERG • LONDON
NEW YORK • OXFORD • PARIS • SAN DIEGO
SAN FRANCISCO • SINGAPORE • SYDNEY • TOKYO
Academic Press is an imprint of Elsevier

Academic Press is an imprint of Elsevier
The Boulevard, Langford Lane, Kidlington, Oxford OX5 1GB, UK
Radarweg 29, PO BOX 211, 1000 AE Amsterdam, The Netherlands
225 Wyman Street, Waltham, MA 02451, USA

First published 2014

British Library Cataloguing in Publication Data
A catalogue record for this book is available from the British Library

Library of Congress Cataloging-in-Publication Data
A catalog record for this book is available from the Library of Congress

ISBN: 978-0-12-417318-7

For information on all Academic Press publications
visit our website at store.elsevier.com

This book has been manufactured using Print On Demand technology. Each copy is produced to order and is limited to black ink. The online version of this book will show color figures where appropriate.

Working together
to grow libraries in
developing countries

www.elsevier.com • www.bookaid.org

CONTENTS

INTRODUCTION

In the first volume, we presented a review and synthesis of our theoretical work, available until now only in journal articles, in which we establish an exact connection between the quantum and classical equations describing the same system. This connection was proved starting from the basic equations, without using any approximation. For stationary multidimensional atomic and molecular systems, the connection between the quantum and classical equations is based on the fact that the geometric elements of the wave described by the Schrödinger equation, namely the wave surface and their normals, are given by the Hamilton–Jacobi equation. For electrodynamic systems composed of a particle in an electromagnetic field, we established a similar connection using the fact that the Klein–Gordon equation is verified exactly by a wave function that corresponds to the classical solution of the relativistic Hamilton–Jacobi equation.

In this volume, we present applications of the above theory, for the modeling of the properties of atomic, molecular, and electrodynamic systems. In the case of the atomic and molecular systems, whose behavior is described by the Schrödinger equation, the principle of our calculation method is based on the fact that the wave function and geometric elements of the wave described by the Schrödinger equation are mathematical objects which describe the same physical system and depend on its constants of motion. It follows that we can use the geometric elements of the wave, to calculate the energetic values and the symmetry properties of the system. Accuracy of our method is comparable to the accuracy of the Hartree–Fock method, for numerous atoms and molecules.

For electrodynamic systems composed of a particle in an electromagnetic field, the connection between the Klein–Gordon and relativistic Hamilton–Jacobi equation is related to a periodicity property, which leads to an accurate method for studying systems composed of very intense laser fields and electrons or atoms. We present a series of applications of this method, such as the calculation of angular and spectral distributions of the radiations generated at interactions between very intense laser beams and electron plasmas or relativistic electron beams. Our results are in good agreement with experimental data from literature. The study of these systems is particularly important, due to the recent emergence of a new generation of ultraintense lasers, whose applications can be predicted using the model described here.

This volume is structured into three chapters, numbered by 1, 2, and 3. Chapters 1 and 2 present wave models for the calculation of the energies and symmetry properties for atomic and molecular systems, while Chapter 3 presents models for the properties of radiations generated at the interaction between very intense laser beams and electron plasmas, electron beams, or atomic gases. For completeness, in Appendices we include the programs which are used for the numerical calculations. The equations are written in the International System.

This book presents applications of the theoretical models, which are described in the first volume, entitled "Theory of Quantum and Classical Connections in Modeling Atomic, Molecular and Electrodynamic Systems." This volume will be referred as "Volume I" throughout the present book.

CHAPTER *1*

Wave Model for Atomic Systems

Abstract

In the first chapter from Volume I, we have shown that the normal curves C of the wave surface of a system described by the Schrödinger equation, can be used to calculate the constants of motion of the system. We show that the motions of the electrons can be separated, and the projection of the C curve from the R^{3N} space of coordinates on the space x_a, y_a, z_a of the electron e_a, denoted by C_a, can be calculated with the aid of the central field approximation. Our method is similar to the central field method, applied in the frame of the atomic and molecular orbitals model. The energy of the system is calculated with the aid of the Bohr quantization relation, which is valid for the C_a curves. The accuracy of our calculations is comparable to the accuracy of the Hartree−Fock method.

Key Words: wave surface; wave surface normals; Hamilton−Jacobi equation; constants of motion; energetic values; Bohr quantization relation; central field approximation; central field method; geometrical symmetries; periodic solutions

1.1 GENERAL CONSIDERATIONS

In this chapter, we review the applications of the wave model presented in the first chapter of Volume I to the case of atomic systems, which were treated in the papers (Popa, 1998b, 1999b, 2008a, 2009a).

The central idea of our approach is that the wave surface and its normals can be used to study the properties of the system, just as well as the wave function. This is because the wave surface and its normals are mathematical elements of the wave described by Schrödinger equation, and depend on the constants of motion of the system. More specifically, we calculate the curves C_a with the aid of the central field method, in a similar manner as the atomic orbital wave functions

Table 1.1 Values of L and $p_{\theta\mathrm{T}}$ for the Atoms Analyzed in This Chapter

Atom	State	Spectral Term	L	$p_{\theta\mathrm{T}}$
Helium	$1s^2$	$^1S_0$	0	0
Lithium	$1s^2 2s$	$^2S_{1/2}$	0	0
Lithium	$1s^2 2p$	$^2P_{1/2}$	1	$\hbar\sqrt{2}$
Beryllium	$1s^2 2s^2$	$^1S_0$	0	0
Boron	$1s^2 2s^2 2p$	$^2P_{1/2}$	1	$\hbar\sqrt{2}$
Carbon	$1s^2 2s^2 2p^2$	$^3P_0$	1	$\hbar\sqrt{2}$
Nitrogen	$1s^2 2s^2 2p^3$	$^4S_{3/2}$	0	0
Oxygen	$1s^2 2s^2 2p^4$	$^3P_2$	1	$\hbar\sqrt{2}$

Ψ_a are calculated in the frame of the atomic orbital method. This similitude explains the fact that the accuracy of our method is comparable to the accuracy of the Hartree–Fock method. In addition, our calculations lead to symmetry properties of the C_a curves, which are similar to the symmetry properties of the wave function.

In virtue of Eq. (1.52) from Volume I, the curve C corresponds to the same constants of motion as those resulting from the Schrödinger equation. For example, the total angular momentum of the curve C is $p_{\theta\mathrm{T}} = \hbar\sqrt{L(L+1)}$, where L is the corresponding quantum number. The value of L can be found from the expression of the spectral term corresponding to a given element, as it results from Table 1.1. Using the data from Landau (1991), in this table we give the spectral terms and the values of L and $p_{\theta\mathrm{T}}$ for all the atoms, which will be analyzed in this chapter.

1.2 SOLUTION FOR HELIUM-LIKE SYSTEMS

The helium-like systems (He, Li^+, Be^{2+}, B^{3+}, and so on) are composed of a nucleus and two 1s electrons, which are denoted by e_1 and e_2. The corresponding C_a curves, for $a = 1, 2$, are the periodic solutions of the equations of motion of the electrons. In a Cartesian system of coordinates with origin at nucleus, these equations are:

$$-\frac{K_1 Z \bar{r}_a}{r_a^3} + \frac{K_1(\bar{r}_a - \bar{r}_b)}{|\bar{r}_a - \bar{r}_b|^3} = m\frac{d^2\bar{r}_a}{dt^2} \quad \text{with} \quad a, b = 1, 2 \text{ and } a \neq b \quad (1.1)$$

where $\bar{r}_a$ is the position vector of the electron e_a and K_1 is given by Eq. (B.3) from Volume I. The system (Eq. (1.1)) has the following solution:

$$\bar{r}_1 = -\bar{r}_2 = \bar{r}, \quad \bar{r}_1 \cdot \bar{k} = \bar{r}_2 \cdot \bar{k} = 0, \quad \bar{v}_1 = -\bar{v}_2 = \bar{v} \tag{1.2}$$

It follows that the motions of the electrons are separated, and the solution of the system (Eq. (1.1)) reduces to the solution of the following equation:

$$-\frac{K_1 Z_a \bar{r}_a}{r_a^3} = m\frac{d^2\bar{r}_a}{dt^2}, \quad \text{where} \quad Z_a = Z - s_{12e} \tag{1.3}$$

and $s_{12e} = s_{21e} = 1/4$. Here, Z_a has the significance of an effective order number while s_{12e} and s_{21e} have the significance of reciprocal screening coefficients of the electrons e_1 and e_2.

In virtue of Eq. (1.3), written for $a = 1$ and $a = 2$, we have

$$E_1 = \frac{m}{2}\left(\frac{dr_1}{dt}\right)^2 - \frac{K_1 Z_1}{r_1}, \quad E_2 = \frac{m}{2}\left(\frac{dr_2}{dt}\right)^2 - \frac{K_1 Z_1}{r_2} \tag{1.4}$$

where E_1 and E_2 are the total energies of the electrons e_1 and e_2. It follows that the total energy is $E = E_1 + E_2$, where, due to the symmetry, $E_1 = E_2$. We note that one half of the electrostatic interaction energy between the two electrons enters in the expression of E_1 through the quantity s_{12e} from the expression of Z_1. The other half of the interaction energy enters in the expression of E_2.

The relation (1.3) is the equation of a motion in central field. It follows that the C_1 curve is an ellipse in the xy plane and, in virtue of Eq. (1.2), the curve C_2 is an ellipse symmetrical to C_1 with respect to the nucleus, as shown in Figure 1.1A.

Since the motions of the electrons are separated, we apply the quantization condition (1.43) from Volume I for the electron e_1 and find its energy, which is E_1. Since its expression is given by Eq. (B.10) from Volume I, we have:

$$E = E_1 + E_2 = 2E_1 = -2R_\infty\frac{Z_1^2}{n_1^2} \tag{1.5}$$

The total angular momentum is:

$$\bar{p}_{\theta T} = \bar{p}_{\theta 1} + \bar{p}_{\theta 2} = 2\bar{p}_{\theta 1} = p_{\theta T}\bar{k} \tag{1.6}$$

where $\bar{p}_{\theta 1}$ and $\bar{p}_{\theta 2}$ are the angular moments corresponding to the curves C_1 and C_2, which are equal.

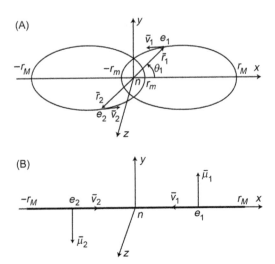

Figure 1.1 *(A) C_1 and C_2 curves for helium-like systems. (B) Quasilinear trajectories for the $1s^2$ states.*

In agreement with the data from Table 1.1, it follows that the angular moments of the C_1 and C_2 curves have negligible values, and we have:

$$p_{\theta 1} = p_{\theta 2} = \varepsilon \quad \text{with} \quad \varepsilon \ll \hbar \tag{1.7}$$

where ε is a very small positive number. It follows that the C_1 and C_2 curves are ellipses with eccentricity e very close to unity, for which the following inequality is strongly fulfilled: $r_m \ll r_M$. These curves are quasilinear along the ox axis, as shown in Figure 1.1B. For these curves, the value of the energy given by Eq. (1.5) must be corrected, taking into account the average spin magnetic interaction energy of the electrons (Gryzinski, 1973):

$$E_{mls} = -\frac{1}{\tau_1} \int_{\tau_1} \bar{\mu}_1 \bar{B}_2 \, dt = -\frac{1}{\tau_1} \int_{\tau_1} \bar{\mu}_2 \bar{B}_1 \, dt \tag{1.8}$$

where τ_1, given by Eq. (B.11) from Appendix B of Volume I, is the period of the electrons motion, while $\bar{\mu}_1$, $\bar{\mu}_2$ and $\bar{B}_1$, $\bar{B}_2$ are, respectively, the magnetic moments and magnetic induction vectors of the two electrons:

$$\mu_1 = \mu_2 = \frac{e\hbar}{2m} \quad \text{and} \quad \bar{B}_2 = \frac{\mu_0}{4\pi} \left[\frac{3\bar{d}(\bar{\mu}_2 \cdot \bar{d})}{d^5} - \frac{\bar{\mu}_2}{d^3} \right] \tag{1.9}$$

a similar relation being valid for $\bar{B}_1$. In the above relations, μ_0 is the magnetic permeability of vacuum and $\bar{d}$ is the vector having its origin on the electron e_2 and the tip on the electron e_1.

In Section 1.3, we prove that E_{m1s} is given by the following relation:

$$E_{m1s} = R_\infty \frac{Z_1^{3/2}}{8n_1^3} \qquad (1.10)$$

We use normalized quantities: the energies are normalized to R_∞ and distances to $2a_0$. The normalized quantities are underlined. For example, $\underline{E}_{m1s} = E_{m1s}/R_\infty$. Introducing the correction term $\underline{E}_{m1s}$ in Eq. (1.5), the normalized expression of the total energy becomes:

$$\underline{E} = 2\underline{E}_1 + \underline{E}_{m1s} = -\frac{2Z_1^2}{n_1^2} + \frac{Z_1^{3/2}}{8n_1^3} \qquad (1.11)$$

where $Z_1 = Z - 1/4$ and $n_1 = 1$.

The experimental value of the total energy, denoted by E_{exp}, is obtained by summing the two ionization energies of helium. The experimental ionization energies are taken from Lide (2003). In Table 1.2, we give a comparison between the experimental value of the total energy of helium and the theoretical values, calculated with the aid of Eq. (1.11), and in papers from literature, with the aid of the Hartree–Fock method (Slater, 1960; Hartree, 1957; Coulson, 1961).

For ions with the same structure as helium, E_{exp} is the sum of the last two ionization energies. The comparison between the theoretical and experimental values of the total energy for helium and for ions with the same structure is presented in Table 1.3.

1.3 EVALUATION OF THE CORRECTION TERM E_{M1S}

The calculation of E_{m1s} is based on the following assumptions:

(a1.1) The vectors $\bar{\mu}_1$ and $\bar{\mu}_2$ are normal to the ox axis, as shown in Figure 1.1B.

(a1.2) The total magnetic moment is equal to zero, corresponding to the $1s^2$ states of helium-like systems.

(a1.3) We suppose that the magnetic forces act only as a perturbation, and it does not change the elliptic character of the C_1 and C_2 curves. When calculating the integral from Eq. (1.8), we approximate these curves by their projections on the xoy plane.

Table 1.2 Normalized Values of the Total Energy Calculated by Popa (2008a, 2009a) and in Literature, with the Aid of the Hartree–Fock Method, as Compared to Experimental Values

Atom	E (Our Model)	E (Literature)	E_{exp}
Helium	−5.83562	−5.723359[a]	−5.80692[b]
		−5.7233598[c]	
		−5.7233600[d]	
Lithium (1s²2s state)	−14.9563	−14.8654514[c]	−14.95634[b]
		−14.8654516[d]	
		−14.8654475[e]	
Lithium (1s²2p state)	−14.8174	−14.760174[f]	−14.82147[g]
		−14.760382[h]	
Beryllium	−29.2533	−29.146044[a]	−29.33766[b]
		−29.146042[c]	
		−29.146042[d]	
Boron	−49.1475	−49.058114[c]	−49.3177[b]
		−49.058116[d]	
		−49.058114[e]	
Carbon	−75.5248	−75.377224[c]	−75.7133[b]
		−75.377232[d]	
		−75.377224[e]	
Nitrogen	− 109.018	−108.80185[c]	−109.2266[b]
		−108.80187[d]	
		108.80184[e]	
Oxygen	−149.428	−149.61874[c]	−150.2206[b]
		−149.61879[d]	
		−149.61876[e]	

The values are given in Rydbergs.
[a]*Huzinaga and Aranu (1970).*
[b]*Lide (2003).*
[c]*Clementi and Roetti (1974).*
[d]*Koga et al. (1993).*
[e]*de Castro and Jorge (2001).*
[f]*Ladner and Goddard (1969).*
[g]*Slater (1960).*
[h]*Beebe and Lunell (1975).*

Assumption (a1.3) is sustained by the fact that the attraction force of the nucleus is overwhelmingly dominant over the most part of the trajectory. We have to check that this force is dominant even in the configuration where the Lorentz force has the maximum value in the vicinity of the nucleus.

Z	2	3	4	5
Table 1.3 Theoretical and Experimental Normalized Values of the Total Energy for Some Atoms and Ions with the Same Structure				
System	He	Li^+	Be^{2+}	B^{3+}
State	$1s^2$	$1s^2$	$1s^2$	$1s^2$
E	-5.83562	-14.555	-27.2173	-43.831
E_{exp} [a]	-5.806921	-14.56004	-27.31391	-44.07097
System	–	Li	Be^+	B^{2+}
State	–	$1s^2 2s$	$1s^2 2s$	$1s^2 2s$
s_{31e}	–	0.854942	0.824677	0.80852
s_{13e}	–	0.0013792	0.0031737	0.0045637
E	–	-14.9563	-28.5522	-46.6073
E_{exp} [a]	–	-14.956336	-28.65244	-46.85889
System	–	Li	Be^+	B^{2+}
State	–	$1s^2 2p$	$1s^2 2p$	$1s^2 2p$
s_{31e}	–	0.979092	0.956543	0.941069
s_{13e}	–	0.000831799	0.00257599	0.00415845
E	–	-14.8174	-28.2684	-46.1839
E_{exp} [b]	–	-14.82147	-28.36230	-46.41877
The values are given in Rydbergs. [a] *Lide (2003).* [b] *Slater (1960).*				

In virtue of Eq. (1.6), we obtain the total orbital magnetic moment:

$$\bar{\mu}_{\theta T} = -\frac{e}{2m}\bar{P}_{\theta T} = -\frac{e}{m}p_{\theta 1}\bar{k} \qquad (1.12)$$

The total magnetic moment is:

$$\bar{\mu}_T = \bar{\mu}_{\theta T} + \bar{\mu}_1 + \bar{\mu}_2 \qquad (1.13)$$

With the aid of assumption (a1.2), we have $\bar{\mu}_T = 0$. We write this relation by components, taking into account Eqs. (1.12), (1.13) and assumption (a1.1), and have $-(e/m)p_{\theta 1}\bar{k} + \mu_{1y}\bar{j} + \mu_{1z}\bar{k} + \mu_{2y}\bar{j} + \mu_{2z}\bar{k} = 0$, from where obtain:

$$\mu_{1z} + \mu_{2z} = \frac{e}{m}p_{\theta 1} \quad \text{and} \quad \mu_{1y} = -\mu_{2y} \qquad (1.14)$$

From relation $\mu_1^2 = \mu_{1y}^2 + \mu_{1z}^2 = \mu_2^2 = \mu_{2y}^2 + \mu_{2z}^2$ and Eq. (1.14), we have $\mu_{1z} = \pm\mu_{2z}$. From Eq. (1.14), we have also $\mu_{1z} + \mu_{2z} > 0$, resulting that $\mu_{1z} = \mu_{2z}$. From this relation, together with Eqs. (1.7) and (1.14), we obtain:

$$\mu_{1z} = \mu_{2z} = \frac{e}{2m} p_{\theta 1} \ll \frac{e}{2m} \hbar \tag{1.15}$$

From the first relation of Eqs. (1.9) and (1.15), we have:

$$\mu_{1z} = \mu_{2z} \ll \mu_1 \tag{1.16}$$

Using Eqs. (1.14) and (1.16), we obtain:

$$\overline{\mu}_1 \cong \mu_1 \overline{j} \quad \text{and} \quad \overline{\mu}_2 \cong -\mu_2 \overline{j} \tag{1.17}$$

as shown in Figure 1.1B.

Figure 1.2A shows the forces which act on the electrons, when the electrons approach the nucleus: the forces in the radial direction, $\overline{F}_{r1}$ and $\overline{F}_{r2}$, the Lorentz forces, $\overline{F}_{m1}$ and $\overline{F}_{m2}$, and the repelling forces between the two electrons, $\overline{F}_{e1}$ and $\overline{F}_{e2}$. It results that the C_1 and C_2 curves are slightly different from the elliptic trajectories described by Eqs. (B.1) and (B.2) from Volume I. In Figure 1.2B, we show the projection of these curves on the plane xy, where γ is the angle between the velocity vector $\overline{v}_1$ and the oy axis, corresponding to $\theta_1 = \pi/2$.

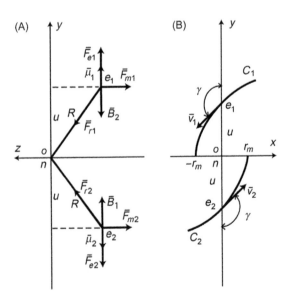

Figure 1.2 (A) Configuration of the electrons in the vicinity of nucleus for $1s^2$ states. (B) Projection of the C_1 and C_2 curves on the xy plane.

The C_1 and C_2 curves have a spatial configuration in the vicinity of the nucleus, where the Lorentz forces cannot be neglected. When the electrons e_1 and e_2 move toward the nucleus, the forces $\bar{F}_{m1}$ and $\bar{F}_{m2}$ act in the direction $-\bar{k}$, as illustrated in Figure 1.2A. It follows that the electrons are situated in the domain corresponding to $z < 0$. When the electrons move away from the nucleus, the forces $\bar{F}_{m1}$ and $\bar{F}_{m2}$ act in the direction $\bar{k}$, and the electrons are situated in the domain for which $z > 0$.

In virtue of assumption (a1.3), we approximate the C_1 and C_2 curves by their projections on the xy plane. From Eq. (1.2), we have $\bar{r}_1 = -\bar{r}_2$. For simplicity, we write $\bar{r}_1 = \bar{r}$ and $\theta_1 = \theta$. From these relations, together with Eqs. (1.8) and (B.1) from Volume I, we have:

$$E_{m1s} = -\frac{1}{\tau_1} \int_{\tau_1} \bar{\mu}_1 \bar{B}_2 \, dt = -\frac{2}{\tau_1} \sqrt{\frac{m}{2|E_1|}} \int_{r_m}^{r_M} \bar{\mu}_1 \bar{B}_2 \frac{r \, dr}{\sqrt{(r_M - r)(r - r_m)}} \quad (1.18)$$

From assumption (a1.3), we can write $\bar{d} \cong 2\bar{r}_1 = 2\bar{r}$. From this relation, together with Eqs. (1.9) and (1.18), we have:

$$E_{m1s} = -\frac{2}{\tau_1} \cdot \frac{\mu_0}{4\pi} \sqrt{\frac{m}{2|E_1|}} \int_{r_m}^{r_M} \left[\frac{12(\bar{r} \cdot \bar{\mu}_1)(\bar{r} \cdot \bar{\mu}_2)}{(2r)^5} - \frac{\bar{\mu}_1 \cdot \bar{\mu}_2}{(2r)^3} \right] \frac{r \, dr}{\sqrt{(r_M - r)(r - r_m)}}$$
$$(1.19)$$

We have $\bar{r} = r \cos\theta \bar{i} + r \sin\theta \bar{j}$, and from Eq. (1.17), we obtain $\bar{r} \cdot \bar{\mu}_1 = r\mu_1 \sin\theta$, $\bar{r} \cdot \bar{\mu}_2 = -r\mu_2 \sin\theta$, and $\bar{\mu}_1 \cdot \bar{\mu}_2 = -\mu_1^2$. On the other hand, from Eq. (B.8) of Volume I, we have $|E_1| = (K_1 Z_1)/r_M$. Introducing these relations in Eq. (1.19), we have:

$$E_{m1s} = \frac{\mu_1^2}{4\tau_1} \frac{\mu_0}{4\pi} \sqrt{\frac{mr_M}{2K_1Z_1}} I \quad \text{where} \quad I = \int_{r_m}^{r_M} \frac{(2 - 3\cos^2\theta)dr}{r^2\sqrt{(r_M - r)(r - r_m)}} \quad (1.20)$$

In Appendix A, we prove that the processing of Eq. (1.20) leads to the expression of the correction term E_{m1s}, which is given by Eq. (1.10).

1.4 SOLUTION FOR LITHIUM-LIKE SYSTEMS

1.4.1 The States $1s^2 2s$

The lithium-like systems (Li, B^+, B^{2+}, C^{3+}, and so on) are composed of a nucleus and three electrons. The 1s electrons are denoted by e_1 and e_2 while the 2s electron is denoted by e_3.

We consider the following assumptions (Gryzinski, 1973; Popa, 2008a):

(a1.4) The motion of the 1s electrons is like in the helium atom, and it is not influenced significantly by the motion of the 2s electron. This assumption is strongly sustained by the fact that the mean distance between a 1s electron and nucleus is much smaller than the mean distance between the 2s electron and nucleus. The C_1 and C_2 curves corresponding to the 1s electrons are quasilinear. In a Cartesian system of coordinates with origin at nucleus (Figure 1.3), we have $r_1 = r_2$, $r_{m1} \cong 0$, and $r_{m2} \cong 0$. They are situated in the plane xoy, very close to the ox axis, as illustrated in Figure 1.3A.

(a1.5) The 2s electron has a quasilinear motion, and the C_3 curve is situated in the plane yoz, very close to the oy axis. The following relation is valid: $r_{m3} \cong 0$. The assumption is sustained by the fact that for this configuration the energy of the system is minimum.

Our solution of the equations of the electron motion is based on a central field method, which has the following stages.

1.4.1.1 Definition of the Screening Coefficients

Since the motions of the e_1 and e_2 electrons are symmetrical, the equation of the energy can be written as:

$$E = -\frac{ZK_1}{r_3} + \frac{m}{2}\left(\frac{dr_3}{dt}\right)^2 + \frac{2K_1}{\sqrt{r_1^2 + r_3^2}} + 2\left[-\frac{(Z - s_{12e})K_1}{r_1} + \frac{m}{2}\left(\frac{dr_1}{dt}\right)^2\right]$$

$$(1.21)$$

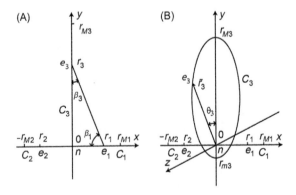

Figure 1.3 Configuration of the curves C_1, C_2, and C_3 for the $1s^2 2s$ (A) and $1s^2 2p$ (B) states of lithium.

We calculate the screening coefficients between the e_1 and e_3 electrons. The force exerted by the electron e_3 on the electron e_1, in the radial direction, is the same as the force exerted by the field of a screening charge, having the value $-es_{13}$, placed on nucleus, where s_{13} is the screening coefficient. The radial components of the forces exerted on the electron e_1 are equal. Using the notations from Figure 1.3A, we can write:

$$\frac{K_1 s_{13}}{r_1^2} = \frac{K_1}{r_1^2 + r_3^2} \cos \beta_1 = \frac{K_1 r_1}{(r_1^2 + r_3^2)^{3/2}} \tag{1.22}$$

In a similar manner, we define the screening coefficient s_{31} as:

$$\frac{K_1 s_{31}}{r_3^2} = \frac{K_1}{r_1^2 + r_3^2} \cos \beta_3 = \frac{K_1 r_3}{(r_1^2 + r_3^2)^{3/2}} \tag{1.23}$$

The ratio between the two screening coefficients is:

$$\frac{s_{13}}{s_{31}} = \frac{r_1^3}{r_3^3} \tag{1.24}$$

The expressions $-K_1 s_{13}/r_1$ and $-K_1 s_{31}/r_3$ are the screening potential energies of the electrons e_1 and e_3. Their sum is equal to the potential energy of the interaction between the electrons e_1 and e_3, because, in virtue of Eqs. (1.22) and (1.23), the following relation is valid:

$$\frac{K_1 s_{13}}{r_1} + \frac{K_1 s_{31}}{r_3} = \frac{K_1 r_1^2}{(r_1^2 + r_3^2)^{3/2}} + \frac{K_1 r_3^2}{(r_1^2 + r_3^2)^{3/2}} = \frac{K_1}{\sqrt{r_1^2 + r_3^2}} \tag{1.25}$$

1.4.1.2 Separation of Variables by Central Field Method

We introduce Eq. (1.25) in Eq. (1.21) and obtain:

$$E = -\frac{(Z - 2s_{31})K_1}{r_3} + \frac{m}{2}\left(\frac{dr_3}{dt}\right)^2 + 2\left[-\frac{(Z - s_{12e} - s_{13})K_1}{r_1} + \frac{m}{2}\left(\frac{dr_1}{dt}\right)^2\right] \tag{1.26}$$

We solve Eq. (1.26) with the aid of a method similar to the self-consistent central field method (Slater, 1960), as follows:

a. We suppose that:

$$s_{31} \cong s_{31e} \quad \text{and} \quad s_{13} \cong s_{13e} \tag{1.27}$$

where s_{31e} and s_{13e} are effective values. In this case, the motion of the system reduces to that in a central field.

b. We recalculate the values of s_{13e} and s_{31e} iteratively, in order to be in agreement with the motion equations.

Similar screening coefficients between the electrons e_2 and e_3, denoted by s_{32e} and s_{23e}, can be defined. Due to the symmetry, we have $s_{13e} = s_{23e}$ and $s_{31e} = s_{32e}$.

From Eqs. (1.26) and (1.27), we have:

$$E = -\frac{Z_3 K_1}{r_3} + \frac{m}{2}\left(\frac{dr_3}{dt}\right)^2 + 2\left[-\frac{Z_1 K_1}{r_1} + \frac{m}{2}\left(\frac{dr_1}{dt}\right)^2\right] \tag{1.28}$$

where

$$Z_1 = Z - s_{12e} - s_{13e} \quad \text{and} \quad Z_3 = Z - 2s_{31e} = Z - 2s_{32e} \tag{1.29}$$

We solve Eq. (1.28) by the separation of the variables method and obtain:

$$E_1 = -\frac{Z_1 K_1}{r_1} + \frac{m}{2}\left(\frac{dr_1}{dt}\right)^2 \quad \text{and} \quad E_3 = -\frac{Z_3 K_1}{r_3} + \frac{m}{2}\left(\frac{dr_3}{dt}\right)^2 \tag{1.30}$$

where E_1 and E_3 are constants having the significance of the total energies of electrons e_1 and e_3. It follows that the total energy is:

$$E = 2E_1 + E_3 \tag{1.31}$$

The equations in (1.30) are identical to those for a hydrogenoid system. It follows that E_1 and E_3 are given by Eq. (B.10) from Volume I:

$$E_1 = -R_\infty \frac{Z_1^2}{n_1^2} \quad \text{and} \quad E_3 = -R_\infty \frac{Z_3^2}{n_1^2} \tag{1.32}$$

where $n_1 = 1$ and $n_3 = 2$, corresponding, respectively, to the states 1s and 2s.

1.4.1.3 Calculation of the Effective Values of the Screening Coefficients

The effective values are evaluated by averaging the basic relations in which they enter. In our case, s_{13e} and s_{31e} are calculated by averaging the equation of the electrical potential of interaction between the two electrons, namely (Eq. (1.25)):

$$s_{13e}\frac{\tilde{1}}{r_1} + s_{31e}\frac{\tilde{1}}{r_3} = \frac{\tilde{1}}{\sqrt{r_1^2 + r_3^2}} \tag{1.33}$$

With the aid of Eqs. (B.1) and (B.11) from Volume I, the first term becomes:

$$\frac{\tilde{1}}{r_1} = \frac{1}{\tau_1} \int_{\tau_1} \frac{dt}{r_1} = \frac{2|E_1|}{n_1\pi\hbar} \sqrt{\frac{m}{2|E_1|}} \int_{r_{m1}}^{r_{M1}} \frac{dr_1}{\sqrt{(r_{M1} - r_1)(r_1 - r_{m1})}} = \frac{\sqrt{2m|E_1|}}{n_1\hbar}$$

(1.34)

Taking into account that C_1 curve is quasilinear, and $r_{m1} \ll r_{M1}$, in virtue of Eqs. (B.10) and (B.12) from Volume I, we have:

$$\frac{\tilde{1}}{r_1} = \frac{1}{n_1\sqrt{ma_0K_1}} \sqrt{2m\frac{Z_1^2K_1}{n_1^2 2a_0}} = \frac{Z_1}{a_0n_1^2} = \frac{1}{a} = \frac{2}{r_{m1} + r_{M1}} \cong \frac{2}{r_{M1}}$$

(1.35)

In a similar manner, we obtain:

$$\frac{\tilde{1}}{r_3} = \frac{1}{\tau_3} \int_{\tau_3} \frac{dt}{r_3} = \frac{2}{r_{m3} + r_{M3}} \cong \frac{2}{r_{M3}}$$

(1.36)

It follows that the values of r_1 and r_3 corresponding to the average positions are:

$$r_{1av} = \frac{r_{M1}}{2} \quad \text{and} \quad r_{3av} = \frac{r_{M3}}{2}$$

(1.37)

Supposing that the 1s electrons are fixed in the average position, we perform the averaging of the interaction term. In a similar manner, with the aid of the relations (B.1) and (B.10)–(B.12) from Volume I, we obtain:

$$\frac{\tilde{1}}{\sqrt{r_1^2 + r_3^2}} = \frac{2}{\pi(r_{m3} + r_{M3})} \int_{r_{m3}}^{r_{M3}} \frac{r_3\,dr_3}{\sqrt{(r_{1av}^2 + r_3^2)(r_{M3} - r_3)(r_3 - r_{m3})}}$$

(1.38)

Since the curve C_3 is quasilinear, we have $r_{m3} \cong 0$, and Eq. (1.38) becomes:

$$\frac{\tilde{1}}{\sqrt{r_1^2 + r_3^2}} = \frac{2}{\pi r_{M3}} \int_0^{r_{M3}} \frac{\sqrt{r_3}\,dr_3}{\sqrt{(r_{1av}^2 + r_3^2)(r_{M3} - r_3)}}$$

(1.39)

From Eqs. (1.33), (1.35)–(1.37), and (1.39), we have:

$$s_{31e} + s_{13e}\frac{r_{M3}}{r_{M1}} = \frac{1}{\pi} \int_0^{r_{M3}} \frac{\sqrt{r_3}\,dr_3}{\sqrt{\left(\frac{r_{M1}^2}{4} + r_3^2\right)(r_{M3} - r_3)}}$$

(1.40)

We write Eq. (1.24) for average positions of the electrons e_1 and e_3, corresponding to r_{1av} and r_{3av}, and obtain:

$$\frac{s_{13e}}{s_{31e}} = \frac{r_{M1}^3}{r_{M3}^3} \tag{1.41}$$

From Eqs. (1.40) and (1.41), we have:

$$s_{31e} = \frac{1}{\pi\left(1 + \frac{r_{M1}^3}{r_{M3}^3}\right)} \int_0^{r_{M3}} \frac{\sqrt{r_3}\, dr_3}{\sqrt{\left(\frac{r_{M1}^2}{4} + r_3^2\right)(r_{M3} - r_3)}} \tag{1.42}$$

We normalize the energies E_1 and E_3 to R_∞ and the distances r_1, r_3, r_{M1}, and r_{M3} to $2a_0$. The normalized quantities are underlined. We consider the relations (1.29), (1.42), and (B.13) from Volume I, for $n_1 = 1$ and $n_3 = 2$, and obtain the following system of three equations with three unknowns, s_{31e}, r_{M1}, and r_{M3}:

$$\underline{r}_{M1} = \frac{n_1^2}{Z_1} \cong \frac{1}{Z - \frac{1}{4}} \tag{1.43}$$

$$\underline{r}_{M3} = \frac{n_3^2}{Z_3} = \frac{1}{Z - 2s_{31e}} \tag{1.44}$$

$$s_{31e} = \frac{1}{\pi\left(1 + \frac{r_{M1}^2}{r_{M3}^2}\right)} \int_0^{\underline{r}_{M3}} \frac{\sqrt{\rho}\, d\rho}{\sqrt{\left(\frac{r_{M1}^2}{4} + \rho^2\right)(\underline{r}_{M3} - \rho)}} \tag{1.45}$$

where, for clarity, we have written $\underline{r}_3 = \rho$.

The system (Eqs. (1.43)–(1.45)) is solved iteratively, according to the following algorithm:

1. Suppose an initial value for s_{31e}, which is denoted s_{31ei}.
2. Calculate $\underline{r}_{M1}$ and $\underline{r}_{M3}$ with the aid of relations (1.43) and (1.44).
3. Recalculate s_{31e} with the aid of relation (1.45), and denote it by s_{31ef}.

If $|1 - s_{31ei}/s_{31ef}| \geq 10^{-6}$, then let $s_{31ei} = s_{31ef}$ and repeat the procedure until the above inequality is not fulfilled. The method converges very quickly; the number of iterations is smaller than 10.

With Eqs. (1.41), (1.43), and (1.44), calculate s_{13e}:

$$s_{13e} = s_{31e} \frac{r_{M1}^3}{\underline{r}_{M3}^3} = \left[\frac{Z - 2s_{31e}}{4\left(Z - \frac{1}{4}\right)}\right]^3 \tag{1.46}$$

1.4.1.4 Calculation of the Total Energy

The total energy is given by Eqs. (1.29), (1.31), and (1.32). We include the correction term given by Eq. (1.10) and obtain:

$$\underline{E} = 2\underline{E}_1 + \underline{E}_{m1s} + \underline{E}_3 = -2\frac{Z_1^2}{n_1^2} + \frac{Z_1^{3/2}}{8n_1^3} - \frac{Z_3^2}{n_3^2} \qquad (1.47)$$

where $Z_1 = Z - 1/4 - s_{13e}$, $Z_3 = Z - 2s_{31e}$, $n_1 = 1$, and $n_3 = 2$.

The experimental value of the total energy of lithium in the fundamental state $1s^2 2s$ is obtained by summing the three ionization energies of lithium, which are taken from Lide (2003). In Table 1.2, we compare the experimental value of the total energy of lithium in the fundamental state with the theoretical values, calculated with the aid of Eq. (1.47), and in a papers from literature, using the Hartree−Fock method. Table 1.4 presents the values of the screening coefficients between the 1s electrons and valence electrons, s_{1ae} and s_{ale}.

For ions with the same structure as lithium in the fundamental state, the experimental values of the total energy are obtained by

Table 1.4 Screening Coefficients Between 1s Electrons and Valence Electrons for the Atoms Analyzed in This Chapter	
Atom	**Screening Coefficients**
Lithium (state $1s^2 2s$)	$s_{31e} = 0.854942$
	$s_{13e} = 0.00137926$
Lithium (state $1s^2 2p$)	$s_{31e} = 0.979092$
	$s_{13e} = 0.000831799$
Beryllium	$s_{31e} = 0.83882$
	$s_{13e} = 0.00221203$
Boron	$s_{31e} = 0.862983$
	$s_{13e} = 0.00246735$
Carbon	$s_{31e} = 0.85505$
	$s_{13e} = 0.00269287$
Nitrogen	$s_{31e} = 0.830382$
	$s_{13e} = 0.0027583$
	$s_{51e} = 0.829599$
	$s_{15e} = 0.00281308$
Oxygen	$s_{31e} = 0.839822$
	$s_{13e} = 0.00284565$

summing the last three ionization energies. The comparison between the theoretical and experimental values of the total energy for lithium and for the ions with the same structure is presented in Table 1.3.

The calculations are made with the aid of a Mathematica 7 program, which is shown in Section B.1.1.

1.4.2 The States 1s²2p

From Table 1.1, we find that $p_{\theta T} = \hbar\sqrt{2}$ for lithium in the state $1s^2 2p$. Since the angular momenta corresponding to the C_1 and C_2 curves have a negligible value, it follows that the angular momentum corresponding to C_3 is:

$$p_{\theta 3} = \hbar\sqrt{2} \tag{1.48}$$

Consequently the curve C_3, corresponding to electron e_3, is an ellipse situated in the plane yoz, while the curves C_1 and C_2, corresponding to the electrons 1s, are quasilinear along the ox axis (Figure 1.3B). The vector $\bar{r}_3$ has the origin on the nucleus and the tip on the curve C_3. The polar coordinates of a point on the ellipse C_3 are $y_3 = r_3 \cos\theta_3$ and $z_3 = r_3 \sin\theta_3$.

The treatment is identical to that presented in Section 1.4.1, and Eqs. (1.21)–(1.32) remain valid.

From Eqs. (1.48) and (B.12) from Volume I, for $n_3 = 2$, we obtain:

$$e = \sqrt{1 - \frac{p_{\theta 3}^2}{n_3^2 \hbar^2}} = \sqrt{0.5} \tag{1.49}$$

and from Eqs. (1.29), (1.49), (B.5), and (B.6) from Volume I, we obtain the following parameters of the ellipse C_3:

$$r_{M3} = \frac{n_3^2 a_0}{Z - 2s_{31e}}(1 + \sqrt{0.5}) \quad \text{and} \quad r_{m3} = \frac{n_3^2 a_0}{Z - 2s_{31e}}(1 - \sqrt{0.5}) \tag{1.50}$$

As in Section 1.4.1, the averaging of the relation of the electrical potential of interaction between the electrons e_1 and e_3, for the $1s^2 2p$ state, leads to an equation identical to Eq. (1.33). The same procedure as in Section 1.4.1 leads to:

$$r_{1av} = \frac{r_{M1}}{2}, \quad r_{3av} = \frac{r_{m3} + r_{M3}}{2}, \quad \frac{s_{13e}}{s_{31e}} = \frac{r_{M1}^3}{(r_{m3} + r_{M3})^3} \tag{1.51}$$

and

$$S_{31e} = \frac{1}{\pi\left[1 + \frac{r_{M1}^2}{(r_{m3}+r_{M3})^2}\right]} \int_{r_{m3}}^{r_{M3}} \frac{r_3 \, dr_3}{\sqrt{\left(\frac{r_{M1}^2}{4} + r_3^2\right)(r_{M3} - r_3)(r_3 - r_{m3})}} \tag{1.52}$$

In this case, we obtain the following system of four equations with four unknowns, s_{31e}, $\underline{r}_{M1}$, $\underline{r}_{M3}$, and $\underline{r}_{m3}$:

$$\underline{r}_{M1} = \frac{n_1^2}{Z_1} \cong \frac{1}{Z - 1/4} \tag{1.53}$$

$$\underline{r}_{M3} = \frac{n_3^2}{2Z_3}(1 + e) = \frac{2}{Z - 2s_{31e}}(1 + \sqrt{0.5}) \tag{1.54}$$

$$\underline{r}_{m3} = \frac{n_3^2}{2Z_3}(1 - e) = \frac{2}{Z - 2s_{31e}}(1 - \sqrt{0.5}) \tag{1.55}$$

$$S_{31e} = \frac{1}{\pi\left[1 + \frac{r_{M1}^2}{(\underline{r}_{m3}+\underline{r}_{M3})^2}\right]} \int_{\underline{r}_{m3}}^{\underline{r}_{M3}} \frac{\rho \, d\rho}{\sqrt{\left(\frac{r_{M1}^2}{4} + \rho^2\right)(\underline{r}_{M3} - \rho)(\rho - \underline{r}_{m3})}} \tag{1.56}$$

where, for clarity, we replaced $\underline{r}_{M3}$ with ρ. The solution of this system is identical to the solution of the system (1.43)–(1.45).

The screening coefficient s_{13e} and the total energy are given by:

$$s_{13e} = s_{31e} \frac{\underline{r}_{M1}^3}{(\underline{r}_{m3}+\underline{r}_{M3})^3} \tag{1.57}$$

$$\underline{E} = 2\underline{E}_1 + \underline{E}_{m1s} + \underline{E}_3 = -2\frac{Z_1^2}{n_1^2} + \frac{Z_1^{3/2}}{8n_1^3} - \frac{Z_3^2}{n_3^2} \tag{1.58}$$

where $Z_1 = Z - 1/4 - s_{13e}$, $Z_3 = Z - 2s_{31e}$, $n_1 = 1$, and $n_3 = 2$.

The experimental value of the total energy of lithium in the state $1s^2 2p$ is taken from Slater (1960, vol. 1, table 15.2). In Table 1.2, we show a comparison between the experimental value and theoretical values, calculated with the aid of Eq. (1.58) and in papers from literature with the aid of the Hartree–Fock method.

The comparison between the theoretical and experimental values (Slater, 1960) of the total energy for lithium in the state $1s^2 2p$ and for ions with the same structure is presented in Table 1.3.

The calculations are made with the aid of a Mathematica 7 program, which is shown in Section B.1.2.

1.5 GEOMETRIC SYMMETRIES AND PERIODIC SOLUTIONS OF THE HAMILTON–JACOBI EQUATION

Our approach from Chapter 1 from Volume I establishes a direct connection between the wave function of the system and the C curves. In this chapter, we show that the wave function and the C curves have the same symmetries.

In a recent paper (Poulsen, 2005), it is shown that the electron configuration that minimizes the energy is the most symmetric configuration allowed by the system. For systems with up to six valence electrons, the symmetric configurations that minimize energy are as follows. For two valence electron systems (like helium and beryllium), the two electrons are diametrically opposite with respect to the nucleus; following Poulsen (2005), this is called the L configuration. For three valence electron systems (like boron), the electrons form an equilateral triangle with the nucleus in center; this is called the ET configuration. For four valence electron systems (like carbon), the electrons are placed at the vertices of a regular tetrahedron; this is called the RTH configuration. Finally, for six valence electron systems (like oxygen), the electrons are placed at the vertices of a regular octahedron with the nucleus in the center; this is called the ROH configuration.

It is shown in Poulsen (2005) that the wave functions of the fundamental states of beryllium, boron, carbon, and oxygen are invariant under the symmetry groups of the L, ET, RTH, and ROH configurations, respectively. We will show below that the C_a curves for valence electrons exhibit the same symmetries as the corresponding wave functions.

We use the same type of screening coefficients, as those defined in Section 1.4, for the study of the valence electrons. In addition, in the general case of two valence electrons, e_a and e_b, situated at equal distances from nucleus, when $r_a = r_b$, a demonstration identical to that from Section 1.4.1 leads to the following formula for the reciprocal screening coefficient of the two electrons:

$$s_{abe} = s_{bae} = \frac{1}{4\sin(\alpha_{ab}/2)} \qquad (1.59)$$

where α_{ab} is the angle between the vectors $\bar{r}_a$ and $\bar{r}_b$ (see also Eq. (3.4) from (Popa, 1999c)).

On the other hand, we consider that the following assumption is valid for valence electrons:

(a1.6) The motion of the 1s electrons is like in the helium atom and the effect of the 1s electrons on the valence electrons is the same as that of a screening charge placed on the nucleus, which has an effective order number $Z' = Z - 2s_{31e}$.

1.6 TYPICAL APPLICATIONS

1.6.1 The Beryllium-Like Systems

The beryllium-like systems (Be, B^+, C^{2+}, N^{3+}, and so on) are composed of a nucleus and four electrons. The 1s electrons are denoted by e_1 and e_2 while the 2s electrons are denoted by e_3 and e_4.

In virtue of assumption (a1.6), the equations of motion for the 2s electrons are:

$$-\frac{K_1 Z' \bar{r}_a}{r_a^3} + \frac{K_1 (\bar{r}_a - \bar{r}_b)}{|\bar{r}_a - \bar{r}_b|^3} = m \frac{d^2 \bar{r}_a}{dt^2} \quad \text{with} \quad a, b = 3, 4 \text{ and } a \neq b \quad (1.60)$$

where $Z' = Z - 2s_{31e}$.

Equation (1.60) is the same as the equation of the 1s electrons for the helium atom, and it has the following solution:

$$\bar{r}_3 + \bar{r}_4 = 0 \qquad (1.61)$$

which in a polar coordinate system in the plane xy, given by $x_a = r_a \cos \theta_a$ and $y_a = r_a \sin \theta_a$, can be written as:

$$r_3 = r_4 \quad \text{and} \quad \theta_4 = \theta_3 + \pi \qquad (1.62)$$

This solution corresponds to an L configuration of the electrons e_3 and e_4, which is illustrated in figure 1a from Poulsen (2005).

The solution of Eq. (1.60) leads to relations identical to Eqs. (1.2) and (1.3), resulting that the curves C_3 and C_4 are two symmetrical quasilinear ellipses, with eccentricities very close to unity. It follows that the energy of electron e_3 is:

$$\underline{E}_3 = -\frac{Z_3^2}{n_3^2} \qquad (1.63)$$

where

$$Z_3 = Z - s_{34e} - 2s_{31e} \tag{1.64}$$

and

$$s_{34e} = s_{12e} = \frac{1}{4} \tag{1.65}$$

Due to the symmetry, we have $\underline{E}_3 = \underline{E}_4$.

The average interaction energy between the magnetic moments of the electrons is given by Eq. (1.10). This relation, written for the 2s electrons, is:

$$E_{m2s} = R_\infty \frac{Z_3^{3/2}}{8n_3^3} \tag{1.66}$$

We suppose that the quasilinear curves C_1 and C_2 are along the ox axis, while C_3 and C_4 are along oy axis. This configuration is similar to that illustrated in Figure 1.3A. Following a procedure identical to that from Section 1.4.1, we find that the screening coefficient s_{31e} can be calculated with the aid of the following system of three equations with three unknowns, s_{31e}, r_{M1}, and r_{M3}:

$$\underline{r}_{M1} = \frac{n_1^2}{Z_1} \simeq \frac{1}{Z - 1/4} \tag{1.67}$$

$$\underline{r}_{M3} = \frac{n_3^2}{Z_3} = \frac{1}{Z - \frac{1}{4} - 2s_{31e}} \tag{1.68}$$

$$s_{31e} = \frac{1}{\pi \left(1 + \frac{r_{M1}^2}{r_{M3}^2} \right)} \int_0^{r_{M3}} \frac{\sqrt{\rho}\, d\rho}{\sqrt{\left(\frac{r_{M1}^2}{4} + \rho^2 \right)(\underline{r}_{M3} - \rho)}} \tag{1.69}$$

As in Section 1.4.1, we obtain the screening coefficient s_{13e} and the total energy:

$$s_{13e} = s_{31e} \frac{r_{M1}^3}{\underline{r}_{M3}^3} \tag{1.70}$$

$$\underline{E} = 2\underline{E}_1 + E_{m1s} + 2\underline{E}_3 + E_{m2s} = -2\frac{Z_1^2}{n_1^2} + \frac{Z_1^{3/2}}{8n_1^3} - 2\frac{Z_3^2}{n_3^2} + \frac{Z_3^{3/2}}{8n_3^3} \tag{1.71}$$

where $Z_1 = Z - 1/4 - 2s_{13e}$, $Z_3 = Z - 1/4 - 2s_{31e}$, $n_1 = 1$, and $n_3 = 2$.

The experimental value of the total energy of beryllium in the fundamental state is obtained by summing the four ionization energies, which are taken from Lide (2003). In Table 1.2, we show a comparison between the experimental value of the total energy of beryllium in the fundamental state and theoretical values, calculated with the aid of Eq. (1.71), and in papers from literature, with the aid of the Hartree–Fock method. The calculations are made with the aid of a Mathematica 7 program, which is shown in Section B.1.3.

1.6.2 The Boron-Like Systems

The boron-like systems (B, C^+, N^{2+}, O^{3+}, and so on) are composed of a nucleus and five electrons. We denote by e_1 and e_2 the 1s electrons, and by e_3, e_4, and e_5 the valence electrons.

In virtue of assumption (a1.6), the equations of motion for valence electrons are:

$$-\frac{K_1 Z' \bar{r}_a}{r_a^3} + \sum_{b,b\neq a} \frac{K_1(\bar{r}_a - \bar{r}_b)}{|\bar{r}_a - \bar{r}_b|^3} = m\frac{d^2\bar{r}_a}{dt^2} \qquad (1.72)$$

where $a, b = 3, 4, 5$, $b \neq a$ and $Z' = Z - 2s_{31e}$.

In this case, the system (Eq. (1.72)) has a solution in the xoy plane. Written in polar coordinates, this solution is given by the following relations:

$$\bar{r}_3 + \bar{r}_4 + \bar{r}_5 = 0 \qquad (1.73)$$

with

$$r_3 = r_4 = r_5 \quad \text{and} \quad \theta_3 = \theta_4 - \frac{2\pi}{3} = \theta_5 - \frac{4\pi}{3} \qquad (1.74)$$

We show that this is a solution of Eq. (1.72), as follows. From Eqs. (1.59), (1.73), and (1.74), we have $s_{34e} = s_{45e} = s_{53e} = 1/(2\sqrt{3})$, $|\bar{r}_3 - \bar{r}_b| = \sqrt{3}r_3$, and $\sum_{b=4,5}(\bar{r}_3 - \bar{r}_b) = 3\bar{r}_3$. Introducing these relations in the expression $\sum_{b,b\neq 3} K_1(\bar{r}_a - \bar{r}_b)/|\bar{r}_a - \bar{r}_b|^3$ which corresponds to $\bar{r}_3$, we have:

$$\sum_{b,b\neq 3} \frac{K_1(\bar{r}_3 - \bar{r}_b)}{|\bar{r}_3 - \bar{r}_b|^3} = \frac{K_1 s_{34e}\bar{r}_3}{r_3^3} + \frac{K_1 s_{35e}\bar{r}_3}{r_3^3} \qquad (1.75)$$

Similar relations are valid for $a = 4, 5$. From Eq. (1.75), it follows that the solution of Eq. (1.72) reduces to the solution of the equation:

$$-\frac{K_1 Z_a \bar{r}_a}{r_a^3} = m\frac{d^2\bar{r}_a}{dt^2} \quad \text{where} \quad a \geq 3 \qquad (1.76)$$

For $a = 3$, we have

$$Z_3 = Z - 2s_{31e} - s_{34e} - s_{35e} = Z - 2s_{31e} - 1/\sqrt{3} \qquad (1.77)$$

Due to symmetry, we have $Z_3 = Z_4 = Z_5$.

Since Eq. (1.76) is the equation of motion in a central field, it results that the curve C_3 is an ellipse in the xy plane. We denote the coordinates of the point P_3, situated at the maximum distances from nucleus, by $P_3(r_{M3}, \theta_{M3})$. In virtue of Eq. (1.74), the curves C_4 and C_5 are ellipses symmetrical to C_3 with respect to the nucleus, having the coordinates of the points situated at the maximum distances of nucleus, respectively, $P_4(r_{M4}, 2\pi/3)$, $P_4(r_{M4}, 4\pi/3)$, where $r_{M3} = r_{M4} = r_{M5}$. The axes of these ellipses are oriented toward the corners of an equilateral triangle, as shown in Figure 1.4, where the distances are normalized at r_{Ma}.

The angular momentum of the curve C_a is:

$$\bar{p}_{\theta a} = m\bar{r}_a \times \frac{d\bar{r}_a}{dt} = mr_a^2 \dot{\theta}_a \bar{k} \qquad (1.78)$$

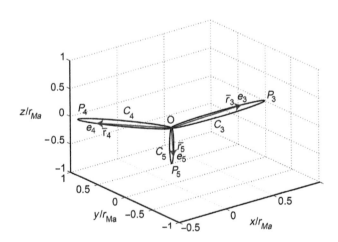

Figure 1.4 Equilateral triangle configuration of the C_a curves of valence electrons in boron atom. The scale is extended in the transversal direction of the ellipses.

From Eqs. (1.74) and (1.78), we have $\dot{\theta}_1 = \dot{\theta}_2 = \dot{\theta}_3$ and $\bar{p}_{\theta 1} = \bar{p}_{\theta 2} = \bar{p}_{\theta 3}$, resulting that the total angular momentum is:

$$p_{\theta T} = \sum_a p_{\theta a} = 3 p_{\theta a} \tag{1.79}$$

From Table 1.1, we find that $p_{\theta T} = \hbar\sqrt{2}$ for boron, and from Eq. (1.79), we obtain:

$$p_{\theta a} = \frac{1}{3} p_{\theta T} = \frac{\sqrt{2}}{3}\hbar \tag{1.80}$$

On the other hand, in virtue of Eq. (1.76), Eq. (1.72) can be solved by separation of variables, and we can apply the relations from Appendix B of Volume I. From Eq. (B.12) from Volume I the eccentricity of the curve C_a can be written as:

$$e = \sqrt{1 - \frac{p_{\theta a}^2}{n_a^2 \hbar^2}} \tag{1.81}$$

where $n_a = 2$.

From Eqs. (1.80) and (1.81), we obtain:

$$e = \sqrt{1 - \frac{1}{18}} \cong 0.97 \tag{1.82}$$

From Eq. (1.77), we have:

$$Z_3 = Z_4 = Z_5 = Z - 2s_{31e} - 2s_{34e} = Z - 2s_{31e} - 1/\sqrt{3} \tag{1.83}$$

The screening coefficient s_{31e} between an 1s electron and a valence electron is calculated with the aid of a system identical to Eqs. (1.53)–(1.55), as follows:

$$r_{M1} = \frac{n_1^2}{Z_1} \cong \frac{1}{Z - 1/4} \tag{1.84}$$

$$r_{M3} = \frac{n_3^2}{2Z_3}(1 + e) = \frac{2}{Z - 2s_{31e} - \frac{1}{\sqrt{3}}}\left(1 + \sqrt{1 - \frac{1}{18}}\right) \tag{1.85}$$

$$r_{m3} = \frac{n_3^2}{2Z_3}(1 - e) = \frac{2}{Z - 2s_{31e} - \frac{1}{\sqrt{3}}}\left(1 - \sqrt{1 - \frac{1}{18}}\right) \tag{1.86}$$

$$s_{31e} = \cfrac{1}{\pi \left[1 + \cfrac{r_{M1}^2}{(r_{m3}+r_{M3})^2}\right]} \int_{r_{m3}}^{r_{M3}} \cfrac{\rho \, d\rho}{\sqrt{\left(\frac{r_{M1}^2}{4} + \rho^2\right)(r_{M3} - \rho)(\rho - r_{m3})}} \tag{1.87}$$

The above system (Eqs. (1.84)–(1.87)) has four equations and four unknowns, s_{31e}, r_{M1}, r_{M3}, and r_{m3}. This system is solved iteratively, according to the algorithm described in Section 1.4.1.

The coefficient s_{13e} is calculated with Eq. (1.57), that is:

$$s_{13e} = s_{31e} \frac{r_{M1}^2}{(r_{m3}+r_{M3})^3} \tag{1.88}$$

Since the electron motions are separated, we can write the total energy as:

$$E = 2\underline{E}_1 + \underline{E}_{m1s} + 3\underline{E}_3 = -2\frac{Z_1^2}{n_1^2} + \frac{Z_1^{3/2}}{8n_1^3} - 3\frac{Z_3^2}{n_3^2} \tag{1.89}$$

where $Z_1 = Z - 1/4 - 3s_{13e}$, $Z_3 = Z - 2s_{31e} - 1/\sqrt{3}$, $n_1 = 1$, and $n_3 = 2$.

The experimental value of the total energy of boron in the fundamental state is obtained by summing the five ionization energies, which are taken from Lide (2003). In Table 1.2, we show a comparison between the experimental value of the total energy of boron in the fundamental state and theoretical values, calculated with the aid of Eq. (1.89), and in papers from literature, with the aid of the Hartree–Fock method. The calculations are made with the aid of a Mathematica 7 program, which is shown in Section B.1.4.

The solution represented by Eqs. (1.73) and (1.74) corresponds to an ET configuration of the electrons e_3, e_4, and e_5, which is illustrated in figure 1b from Poulsen (2005).

1.6.3 The Carbon-Like Systems

The carbon-like systems (C, N^+, O^{2+}, and so on) are composed of a nucleus and six electrons. We denote by e_1 and e_2 the 1s electrons, and by e_3, e_4, e_5, and e_6 the valence electrons.

The C_a curves for the valence electrons of carbon are given by the system (Eq. (1.72)), where $a, b = 3, 4, 5, 6$, $a \neq b$ and $Z' = Z - 2s_{31e}$.

We consider a cylindrical system of coordinates, where the radial, azimuthal, and vertical coordinates are denoted, respectively, by ρ, θ, and z, resulting that the position vector of an electron is:

$$\bar{r}_a = \bar{\rho}_a + z_a \bar{k} = \rho_a \cos \theta_a \bar{i} + \rho_a \sin \theta_a \bar{j} + z_a \bar{k} \tag{1.90}$$

The system (Eq. (1.72)) has the following periodical solution:

$$\bar{r}_3 + \bar{r}_4 + \bar{r}_5 + \bar{r}_6 = 0 \quad \text{with} \quad r_3 = r_4 = r_5 = r_6 \tag{1.91}$$

$$\rho_3 = \rho_4 = \rho_5 = \rho_6 = r_a \cos \alpha, \quad \text{where} \quad a = 3, 4, 5, 6 \tag{1.92}$$

$$\cos \alpha = \sqrt{\frac{2}{3}} \quad \text{and} \quad \theta_3 = \theta_4 - \frac{\pi}{2} = \theta_5 - \pi = \theta_6 - \frac{3\pi}{2} \tag{1.93}$$

$$z_3 = z_5 = r_a \sin \alpha = \frac{1}{\sqrt{3}} r_a \quad \text{and} \quad z_4 = z_6 = -r_a \sin \alpha = -\frac{1}{\sqrt{3}} r_a \tag{1.94}$$

where α is the angle between the vectors $\bar{r}_a$ and $\bar{\rho}_a$.

We show that the relations (1.91)–(1.94) are a solution of Eq. (1.72), as follows. From Eqs. (1.59) and (1.91)–(1.94), we have $s_{34e} = s_{35e} = s_{36e} = \sqrt{3}/(4\sqrt{2})$, $|\bar{r}_3 - \bar{r}_b| = 2\sqrt{2/3}r_3$ with $b = 4, 5, 6$ and $\sum_{b,b\neq3}(\bar{r}_3 - \bar{r}_b) = 4\bar{r}_3$. Introducing these relations in the expression $\sum_{b,b\neq3} K_1(\bar{r}_a - \bar{r}_b)/|\bar{r}_a - \bar{r}_b|^3$ which corresponds to $\bar{r}_3$, we have:

$$\sum_{b,b\neq3} \frac{K_1(\bar{r}_3 - \bar{r}_b)}{|\bar{r}_3 - \bar{r}_b|^3} = \frac{K_1 s_{34e} \bar{r}_3}{r_3^3} + \frac{K_1 s_{35e} \bar{r}_3}{r_3^3} + \frac{K_1 s_{36e} \bar{r}_3}{r_3^3} \tag{1.95}$$

Similar relations are valid for $a = 4, 5, 6$. From Eq. (1.95), it follows that the solution of Eq. (1.72) reduces to the solution of Eq. (1.76). The order number from Eq. (1.76), for $a = 3$, is

$$Z_3 = Z - 2s_{31e} - 3s_{34e} = Z - 2s_{31e} - 3\sqrt{3}/(4\sqrt{2}) \tag{1.96}$$

Due to symmetry, we have $Z_3 = Z_4 = Z_5 = Z_6$.

The following relation is valid in cylindrical coordinates (Arfken, 1985):

$$\frac{d^2\bar{r}_a}{dt^2} = \left(\frac{d^2\rho_a}{dt^2} - \rho_a \dot{\theta}_a^2\right)\hat{\rho}_a + \frac{1}{\rho_a}\frac{d}{dt}(\rho_a^2 \dot{\theta}_a)\hat{\theta}_a + \frac{d^2 z_a}{dt^2}\hat{z}_a \tag{1.97}$$

where $\hat{\rho}_a$, $\hat{\theta}_a$, and $\hat{z}_a$ are the versors of the cylindrical coordinates, and $\hat{z}_a \equiv \overline{k}$. In virtue of Eqs. (1.90) and (1.97), Eq. (1.76) is equivalent to the following three equations:

$$m\left(\frac{d^2\rho_3}{dt^2} - \rho_3\dot{\theta}_3^2\right) = -\frac{K_1 Z_3 \cos^3 \alpha}{\rho_3^2} \tag{1.98}$$

$$m\rho_3^2\dot{\theta}_3 = p_{\theta 3z} = \text{constant} \tag{1.99}$$

$$m\frac{d^2 z_3}{dt^2} = -\frac{K_1 Z_3 \sin^3 \alpha}{z_3^2} \tag{1.100}$$

Taking into account Eqs. (1.99) and (1.100), it follows that the component of the electron motion on the xy plane is elliptical, having angular momentum equal to $p_{\theta 3z}$. We denote by C_{3xy} the ellipse resulting from Eqs. (1.99) and (1.100). From Eq. (1.93), it follows that the four ellipses C_{3xy}, C_{4xy}, C_{5xy}, and C_{6xy} are symmetrical (Figure 1.5).

We assume that the ellipses have the eccentricity very close to unity. In virtue of the theory of the elliptical motion (Landau and Lifschitz, 2000), in this case the term $\rho_3\dot{\theta}_3^2$ is negligible, and Eq. (1.100) becomes:

$$m\frac{d^2\rho_3}{dt^2} \cong -\frac{K_1 Z_3 \cos^3 \alpha}{\rho_3^2} \tag{1.101}$$

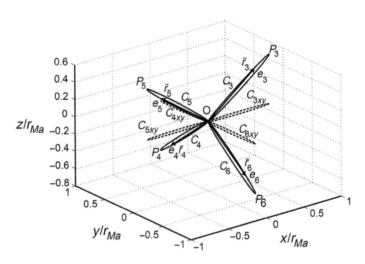

Figure 1.5 Tetrahedron configuration of the C_a curves of valence electrons in carbon atom. The scale is extended in the transversal direction of the ellipses. The projections of the C_a curves on the xy plane are represented by dotted lines.

To justify this assumption, we will calculate below the eccentricity of the curve C_{3xy}.

The vector $\bar{r}_3$ makes a constant angle α with the xy plane. It follows that the curve C_3 is situated on a half cone, whose vertex is in the origin of the coordinate axes, its axis is the oz coordinate axis and the opening angle equal to $2(\pi/2 - \alpha)$. The projection of the curve C_3 onto the xy plane is the curve C_{3xy} (see Figure 1.5). The curve C_5 can be described in the same way, while the curves C_4 and C_6 lie on a symmetrical half cone with respect to the nucleus, such that their projections onto the xy plane are the curves C_{4xy} and C_{6xy}.

The curves $C_3 - C_6$ have transversal dimensions very small compared to their lengths, resulting that they are, to very good approximation, planar curves. Since their projections onto the xy plane are ellipses, it results that $C_3 - C_6$ are also, with good approximations, ellipses with eccentricities very close to unity.

We denote the coordinates of the point P_3 situated on the curve C_3, at the maximum distances from nucleus, by $P_3(\rho_{M3}, \theta_{M3}, z_{M3})$, and choose $\theta_{M3} = 0$. In virtue of Eqs. (1.90)–(1.94), the curves C_4, C_5, and C_6 are symmetrical to C_3 with respect to the nucleus, having the coordinates of the points situated at the maximum distances of nucleus, respectively, $P_4(\rho_{M4}, \pi/2, z_{M4})$, $P_5(\rho_{M5}, \pi, z_{M5})$, and $P_6(\rho_{M6}, 3\pi/2, z_{M6})$, where $\rho_{M3} = \rho_{M4} = \rho_{M5} = \rho_{M6}$ and $|z_{M3}| = |z_{M4}| = |z_{M5}| = |z_{M6}|$. The points P_3, P_4, P_5, and P_6 are situated on the vertices of a regular tetrahedron, as shown in Figure 1.5, where the distances are normalized at maximum distances from nucleus, denoted by r_{Ma}.

The angular momentum of the curve C_a is (Arfken, 1985):

$$\bar{p}_{\theta a} = m\bar{r}_a \times \dot{\bar{r}}_a = m(\rho_a \hat{\rho}_a + z_a \hat{z}_a) \times (\dot{\rho}_a \hat{\rho}_a + \rho_a \dot{\theta}_a \hat{\theta}_a + \dot{z}_a \hat{z}_a)$$
$$= m[-z_a \rho_a \dot{\theta}_a \hat{\rho}_a + (-\rho_a \dot{z}_a + \dot{\rho}_a z_a)\hat{\theta}_a + \rho_a^2 \dot{\theta}_a \hat{z}_a] \tag{1.102}$$

From Eqs. (1.92) and (1.94), we have $|z_a| = \rho_a \tan \alpha$, resulting that the absolute value of the angular momentum of the curve C_a is:

$$p_{\theta a} = m\frac{\rho_a^2 \dot{\theta}_a}{\cos \alpha} = mr_a^2 \dot{\theta}_a \cos \alpha \tag{1.103}$$

The projection of the vector $\bar{p}_{\theta a}$ on the oz axis is:

$$p_{\theta az} = m\rho_a^2 \dot{\theta}_a = mr_a^2 \dot{\theta}_a \cos^2 \alpha = p_{\theta a} \cos \alpha \tag{1.104}$$

From Eqs. (1.92) and (1.93), we have $\dot{\theta}_1 = \dot{\theta}_2 = \dot{\theta}_3 = \dot{\theta}_4$ and $\cos \alpha = \sqrt{2/3}$, resulting that $\overline{p}_{\theta 1} = \overline{p}_{\theta 2} = \overline{p}_{\theta 3} = \overline{p}_{\theta 4}$ and $p_{\theta 1z} = p_{\theta 2z} = p_{\theta 3z} = p_{\theta 4z}$. The total angular momentum is:

$$p_{\theta T} = \sum_a \overline{p}_{\theta az} = \sum_a \overline{p}_{\theta a} \cos \alpha = 4\sqrt{\frac{2}{3}} p_{\theta a} \qquad (1.105)$$

From Table 1.1, we find that $p_{\theta T} = \hbar\sqrt{2}$ in the case of carbon. From Eq. (1.105), we obtain:

$$p_{\theta a} = \frac{\sqrt{3}}{4\sqrt{2}} p_{\theta T} = \frac{\sqrt{3}}{4} \hbar \qquad (1.106)$$

and from Eqs. (1.81) and (1.106), we have:

$$e = \sqrt{1 - \frac{3}{64}} \cong 0.98 \qquad (1.107)$$

This value is very close to unity. On the other hand, in virtue of Eq. (1.104), we have $p_{\theta az} < p_{\theta a}$, and from Eq. (1.81), it follows that the eccentricity of the ellipse C_{axy} is more closer to unity than e, and the assumption made before Eq. (1.101) is verified.

From Eq. (1.96), we have:

$$Z_3 = Z_4 = Z_5 = Z_6 = Z - 2s_{31e} - 3s_{34e} = Z - 2s_{31e} - \frac{3\sqrt{3}}{4\sqrt{2}} \qquad (1.108)$$

The screening coefficients s_{31e} and s_{13e} are calculated with the aid of Eqs. (1.84)–(1.88), where we substitute the values of e and Z_3 given, respectively, by Eqs. (1.107) and (1.108):

Since the electron motions are separated, we can write the total energy as:

$$\underline{E} = 2\underline{E}_1 + \underline{E}_{m1s} + 4\underline{E}_3 = -2\frac{Z_1^2}{n_1^2} + \frac{Z_1^{3/2}}{8n_1^3} - 4\frac{Z_3^2}{n_3^2} \qquad (1.109)$$

where $Z_1 = Z - 1/4 - 4s_{13e}$, $Z_3 = Z - 2s_{31e} - 3\sqrt{3}/4\sqrt{2}$, $n_1 = 1$, and $n_3 = 2$.

The experimental value of the total energy of carbon in the fundamental state is obtained by summing the six ionization energies, which are taken from Lide (2003). In Table 1.2, we show a comparison between the experimental value of the total energy of carbon in the fundamental state and theoretical values, calculated with the aid of Eq. (1.109), and in papers from literature, with the aid of the Hartree−Fock method. The calculations are made with the aid of a Mathematica 7 program, which is shown in Section B.1.5.

The solution (Eqs. (1.91)−(1.94)) corresponds to an RTH configuration of the valence electrons, in which the center of the coordinate system is at the center of a tetrahedron, and the four electrons are situated at four vertices of the tetrahedron. This configuration is illustrated in figure 1c from Poulsen (2005).

1.6.4 The Oxygen-Like Systems

The oxygen-like systems (O, F^+, Ne^{2+}, and so on) are composed of a nucleus and eight electrons. We denote by e_1 and e_2 the 1s electrons, and by $e_3, e_4, \ldots, e_8$ the valence electrons.

The C_a curves for the valence electrons of oxygen are given by the system (Eq. (1.72)), where $a, b = 3, 4, \ldots, 8$, $b \neq a$ and $Z' = Z - 2s_{31e}$.

We consider a cylindrical system of coordinates, ρ, θ, and z, as in Section 1.6.3. The system (Eq. (1.72)) has a solution similar to Eqs. (1.91)−(1.94), as follows:

$$\bar{r}_3 + \bar{r}_4 + \cdots + \bar{r}_8 = 0 \quad \text{with} \quad r_3 = r_4 = \cdots = r_8 \tag{1.110}$$

$$\rho_3 = \rho_4 = \cdots = \rho_8 = r_a \cos \alpha, \quad \text{where} \quad a = 3, 4, \ldots, 8 \tag{1.111}$$

$$\cos \alpha = \sqrt{\frac{2}{3}} \quad \text{and} \quad \theta_3 = \theta_4 - \frac{\pi}{3} = \theta_5 - \frac{2\pi}{3} = \cdots = \theta_8 - \frac{5\pi}{3} \tag{1.112}$$

$$z_3 = z_5 = z_7 = r_a \sin \alpha = \frac{1}{\sqrt{3}} r_a \tag{1.113}$$

$$z_4 = z_6 = z_8 = -r_a \sin \alpha = -\frac{1}{\sqrt{3}} r_a \tag{1.114}$$

We show that the relations (1.110)−(1.114) are a solution of Eq. (1.72), as follows. From Eqs. (1.59) and (1.110)−(1.114), we have $s_{34e} = s_{35e} = s_{37e} = s_{38e} = 1/(2\sqrt{2})$, $s_{36e} = 1/4$, $|\bar{r}_3 - \bar{r}_4| = |\bar{r}_3 - \bar{r}_5| = |\bar{r}_3 - \bar{r}_7| = |\bar{r}_3 - \bar{r}_8| = \sqrt{2}\, r_3$, $|\bar{r}_3 - \bar{r}_6| = 2\, r_3$, $\sum_{b=4,5,7,8}(\bar{r}_3 - \bar{r}_b) = 4\bar{r}_3$ and

$\bar{r}_3 - \bar{r}_6 = 2\bar{r}_3$. Introducing these relations in the expression $\sum_{b,b\neq 3}$ $K_1(\bar{r}_a - \bar{r}_b)/|\bar{r}_a - \bar{r}_b|^3$ which corresponds to $\bar{r}_3$, we have:

$$\sum_{b,b\neq 3} \frac{K_1(\bar{r}_3 - \bar{r}_b)}{|\bar{r}_3 - \bar{r}_b|^3} = \frac{K_1(s_{34e} + s_{35e} + s_{36e} + s_{37e} + s_{38e})\bar{r}_3}{r_3^3} \qquad (1.115)$$

Similar relations are valid for $a = 4, 5, \ldots, 8$. From Eq. (1.115), it follows that the solution of Eq. (1.72) reduces to the solution of Eq. (1.76).

The expression of the order number which enters in Eq. (1.76), written for $a = 3$, is

$$Z_3 = Z - 2s_{31e} - 4s_{34e} - s_{36e} = Z - 2s_{31e} - \sqrt{2} - 1/4 \qquad (1.116)$$

Due to symmetry, we have $Z_3 = Z_4 = \cdots = Z_8$.

From this point, the analysis is identical to that from Section 1.6.3, and the Eqs. (1.97)–(1.104) remain valid. We find that the C_a curves are six ellipses with the eccentricities very close to unity. We denote the coordinates of the point P_3 situated on the curve C_3, at the maximum distance from nucleus, by $P_3(\rho_{M3}, \theta_{M3}, z_{M3})$, and choose $\theta_{M3} = 0$. The curves $C_4, C_5, \ldots, C_8$ are symmetrical to C_3, having the coordinates of the points situated at the maximum distances of nucleus, respectively, $P_4(\rho_{M4}, \pi/3, z_{M4})$, $P_5(\rho_{M5}, 2\pi/3, z_{M5})$, $\ldots$, $P_8(\rho_{M8}, 5\pi/3, z_{M8})$, where $\rho_{M3} = \rho_{M4} = \cdots = \rho_{M8}$ and $|z_{M3}| = |z_{M4}| = \cdots = |z_{M8}|$. The points P_3, $P_4, \ldots, P_8$ are situated on the vertices of a regular octahedron, as shown in Figure 1.6, where the distances are normalized at maximum distances from nucleus.

Now we check that the eccentricity of the curves C_a is very close to unity. From Table 1.1 we find that $p_{\theta T} = \hbar\sqrt{2}$ in the case of oxygen. Since there are six valence electrons, Eq. (1.105) can be written as follows:

$$p_{\theta T} = \sum_{\alpha} p_{\theta az} = \sum_{a} p_{\theta a} \cos \alpha = 6\sqrt{\frac{2}{3}} p_{\theta a} \qquad (1.117)$$

and the angular momentum of the curves $C_3 - C_8$ is equal to:

$$p_{\theta a} = \frac{\sqrt{3}}{6\sqrt{2}} p_{\theta T} = \frac{\sqrt{3}}{6} \hbar \qquad (1.118)$$

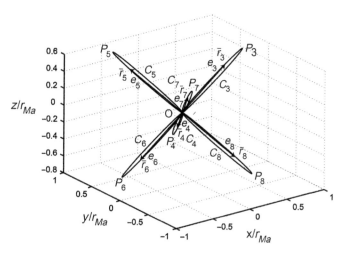

Figure 1.6 Octahedron configuration of the C_a curves of valence electrons in oxygen atom. The scale is extended in the transversal direction of the ellipses.

From Eqs. (1.81) and (1.118), we have:

$$e = \sqrt{1 - \frac{3}{144}} \cong 0.99 \tag{1.119}$$

This value is very close to unity. We can show, as in Section 1.6.3, that the eccentricity of the ellipse C_{axy} is more closer to unity than e, and the assumption made before Eq. (1.101) is verified.

From Eq. (1.116), we have:

$$Z_3 = Z_4 = \cdots = Z_8 = Z - 2s_{31e} - \sqrt{2} - \frac{1}{4} \tag{1.120}$$

The screening coefficients s_{31e} and s_{13e} are calculated with the aid of Eqs. (1.84)–(1.88), where we substitute the values of e and Z_3 given respectively, by Eqs. (1.119) and (1.120).

Since the electron motions are separated, we can write the total energy as:

$$\underline{E} = 2\underline{E}_1 + \underline{E}_{mls} + 6\underline{E}_3 = -2\frac{Z_1^2}{n_1^2} + \frac{Z_1^{3/2}}{8n_1^3} - 6\frac{Z_3^2}{n_3^2} \tag{1.121}$$

where $Z_1 = Z - 1/4 - 6s_{13e}$, $Z_3 = Z - 2s_{31e} - \sqrt{2} - 1/4$, $n_1 = 1$, and $n_3 = 2$.

The experimental value of the total energy of oxygen in the fundamental state is obtained by summing the eight ionization energies, which are taken from Lide (2003). In Table 1.2, we show a comparison between the experimental value of the total energy of oxygen in the fundamental state and theoretical values, calculated with the aid of Eq. (1.121), and in papers from literature, with the aid of the Hartree−Fock method. The calculations are made with the aid of a Mathematica 7 program, which is shown in Section B.1.6.

The solution (Eqs. (1.110)−(1.114)) corresponds to an ROH configuration of the valence electrons, in which the center of the coordinate system is a center of a cube, and the six electrons are situated at the centers of the six faces of the cube. This configuration is illustrated in figure 1d from Poulsen (2005).

In conclusion, the wave functions of the fundamental states have the same symmetry as the C_a curves, in the case of beryllium, boron, carbon, and oxygen atoms.

1.7 A MORE GENERAL METHOD APPLIED TO THE NITROGEN ATOM

It is shown in Poulsen (2005) that the energetically most favorable geometric configuration, for a multielectron system corresponds to the maximum symmetry of the geometrical configuration of the electrons. The same symmetry was proved for the C_a curves in previous sections. Moreover, the solutions of the equations of the C_a curves (Eqs. (1.61), (1.62), (1.90)−(1.94), and so on) correspond to the case when the averaged positions of the electrons are symmetrical.

We suppose that in the general case of multielectron stationary system the average positions of the electrons are symmetrical. The average distance of the electron from the nucleus is given by Eq. (B.14) from Volume I.

We show that the above assumption leads to a strong simplification of the calculation of the total energy, with application to the case of nitrogen atom. The nitrogen atom is composed of a nucleus and seven electrons. We denote by e_1 and e_2 the 1s electrons, and by e_3, e_4, ..., e_7 the valence electrons. The averaged positions of the valence electrons of the nitrogen, which correspond to the maximum symmetry of the

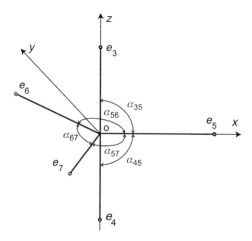

Figure 1.7 Symmetrical average positions of the valence electrons in the nitrogen atom.

system, are shown in Figure 1.7. Corresponding to these positions, which are denoted by $e_3, e_4, \ldots, e_7$, we have $\alpha_{34} = \pi$, $\alpha_{35} = \alpha_{36} = \alpha_{37} = \pi/2$, and $\alpha_{56} = \alpha_{57} = \alpha_{67} = 2\pi/3$. In virtue of Eq. (1.59), the reciprocal screening coefficients of the valence electrons results as follows:

$$s_{34e} = \frac{1}{4}, \quad s_{35e} = s_{36e} = s_{37e} = \frac{1}{2\sqrt{2}} \tag{1.122}$$

and

$$s_{56e} = s_{57e} = s_{67e} = \frac{1}{2\sqrt{3}} \tag{1.123}$$

Due to symmetry, we have $s_{abe} = s_{bae}$, and the effective order numbers of the valence electrons can be written as:

$$Z_3 = Z_4 = Z - 2s_{31e} - s_{34e} - 3s_{35e} = Z - 2s_{31e} - \frac{1}{4} - \frac{3}{2\sqrt{2}} \tag{1.124}$$

and

$$Z_5 = Z_6 = Z_7 = Z - 2s_{51e} - 2s_{35e} - 2s_{56e}$$
$$= Z - 2s_{51e} - \frac{1}{\sqrt{2}} - \frac{1}{\sqrt{3}} \tag{1.125}$$

On the other hand, from Table 1.1 we find that $p_{\theta T} = 0$ in the case of the fundamental state of the nitrogen, resulting that the C_a curves can be approximated by ellipses with negligible minimum distance from nucleus. In this case, the screening coefficients s_{31e} and s_{13e} can be calculated with the relations (1.67)–(1.70), where the expression of Z_3 is given by Eq. (1.124).

For the calculation of the screening coefficients s_{51e} and s_{15e}, we use again relations similar to Eqs. (1.67)–(1.70).

The total energy results, as follows:

$$E = 2E_1 + E_{m1s} + 2E_3 + 3E_5 = -2\frac{Z_1^2}{n_1^2} + \frac{Z_1^{3/2}}{8n_1^3} - 2\frac{Z_3^2}{n_3^2} + 3\frac{Z_5^2}{n_5^2} \quad (1.126)$$

where $Z_1 = Z - 1/4 - 2s_{13e} - 3s_{15e}$, $Z_3 = Z - 2s_{31e} - 1/4 - 3/(2\sqrt{2})$, $Z_5 = Z - 2s_{51e} - 1/\sqrt{2} - 1/\sqrt{3}$, $n_1 = 1$, and $n_3 = n_5 = 2$.

The experimental value of the total energy of nitrogen in the fundamental state is obtained by summing the seven ionization energies, which are taken from Lide (2003). In Table 1.2, we show a comparison between the experimental value of the total energy of nitrogen in the fundamental state and theoretical values, calculated with the aid of Eq. (1.126), and in papers from literature, with the aid of the Hartree–Fock method. The calculations are made with the aid of a Mathematica 7 program, which is shown in Section B.1.7.

1.8 GENERAL RELATIONS DERIVED FOR THE CENTRAL FIELD METHOD

The main point of the method presented in this chapter is that the central field approximation makes possible the separation of the variables and the calculations of the C_a curves. In this case, the atomic systems are described by the following relations:

$$E = E_1 + E_2 + \sum_{a \geq 3} E_a \quad (1.127)$$

$$E_1 = E_2 = -\frac{K_1 Z_1}{r_1} + \frac{m}{2}\left(\frac{dr_1}{dt}\right)^2 = -\frac{R_\infty Z_1^2}{n_1^2} \quad (1.128)$$

$$E_a = -\frac{K_1 Z_a}{r_a} + \frac{m}{2}\left(\frac{dr_a}{dt}\right)^2 = -\frac{R_\infty Z_a^2}{n_a^2} \quad \text{for} \quad a \geq 3 \quad (1.129)$$

$$Z_1 = Z_2 = Z - s_{12e} - \sum_{a \geq 3} s_{1ae} \quad \text{with} \quad s_{12e} = \frac{1}{4} \quad (1.130)$$

$$Z_a = Z - 2s_{a1e} - \sum_{b \geq 3; b \neq a} s_{abe} \quad \text{for} \quad a \geq 3 \qquad (1.131)$$

where s_{a1e} and s_{1ae} are calculated with the relations (1.67)–(1.70) or (1.84)–(1.88), while s_{abe}, for $a, b \geq 3$, is calculated with the relation (1.59).

In Chapter 2, we will show that similar relations can be generalized in the case of the molecules.

Wave Model for Molecular Systems

Abstract

The central field method for the calculation of atom properties can be extended to the case of diatomic molecules. More specifically, the C_a curves, which are elements of the wave described by the Schrödinger equation, can be calculated in the case of diatomic molecules. Using the properties of the C_a curves, we calculate the energetic values and symmetry properties of diatomic molecules, and we give examples for the molecules Li_2, Be_2, B_2, C_2, LiH, BeH, BH, and CH. As in the case of atoms, the accuracy of the method is comparable to the accuracy of the Hartree–Fock method, for the same system. Our approach also explains basic properties of the molecules in discussion.

Keywords: homonuclear diatomic molecules; heteronuclear diatomic molecules; energetic values; symmetry properties; central field method; covalent bond; single; double and triple bonds; ionic bond; electric dipole moment.

2.1 GENERAL CONSIDERATIONS

In this chapter, we review the applications of the wave model presented in the first chapter from Volume I to the case of molecular systems, which were treated in the papers (Popa, 1999b,c, 2000, 2011a).

We use symmetry properties of the systems to show that the central field method for calculating the energetic values, presented in Chapter 1, can be extended to diatomic molecules, both homonuclear and heteronuclear, with the same accuracy as that of the Hartree–Fock method. We compare our data with theoretical data taken from the Computational Chemistry Comparison and Benchmark Database (see Ref. (CCCBDB)). In addition, the method presented in this paper predicts basic properties of the molecules (Coulson, 1961; Pauling, 1970), such as the existence of the single, double, and triple bonds in the case of the C−C bond, the symmetry properties of these bonds, the fact that in the case of the C_2 molecule the bond is double, and it also explains the ionic character of the LiH bond.

The symbols used in the equations that describe the behavior of the molecules contain supplementary indexes, which refer to different nuclei of the molecule. These indexes are as follows:

1. The 1s electrons which move in the field of the nucleus n_A are denoted by e_{A1}, e_{A2}, while the valence electrons which move in the field of the nucleus n_A and do not participate to the bond are denoted by e_{A3}, e_{A4}, e_{A5}, and so on. Similar notations correspond to electrons which move in the field of the nucleus n_B. The bond electrons are denoted by e_1, e_2, e_3, and so on.
2. In the case of the homonuclear diatomic molecules, the nuclei order numbers, denoted by Z_A and Z_B, are equal. In this case, the order numbers corresponding to the electrons e_{Ai} and e_j, where $i,j = 1, 2, \ldots$, are denoted, respectively, by Z_{Ai} and Z_j. The screening coefficients between electrons e_{Ai} and e_{Ak} are denoted by $s_{Ai,Ak}$ and $s_{Ak,Ai}$, those between the bond electrons e_j and e_k are denoted by $s_{j,k}$ and $s_{k,j}$, and the screening coefficients between the electrons e_{Ai} and e_j are denoted by $s_{Ai,j}$ and $s_{j,Ai}$.
3. The effective order number of the nucleus n_A in interactions between nuclei includes the effect of the 1s electrons and the effect of the electrons that do not participate to the bond. In this case, the effective order number of n_A is denoted by Z_{nA}.

2.2 CALCULATIONS OF THE C_A CURVES CORRESPONDING TO SINGLE, DOUBLE, AND TRIPLE BONDS OF HOMONUCLEAR MOLECULES AND TO IONIC AND COVALENT BONDS OF HETERONUCLEAR MOLECULES

2.2.1 C_a Curves for Single, Double, and Triple Bonds in Homonuclear Molecules

We consider a molecule composed of two fixed identical nuclei, denoted by n_A and n_B, two 1s electrons in the vicinity of the n_A nucleus, denoted by e_{A1} and e_{A2}, two 1s electrons in the vicinity of the n_B nucleus, denoted by e_{B1} and e_{B2}, and N_b valence electrons which participate to the bond, denoted by e_1, e_2, $\ldots$, e_{Nb}. The Cartesian coordinates of the nuclei are $n_A(-\sigma, 0, 0)$ and $n_B(\sigma, 0, 0)$, and their order number is equal to Z_A. The equations of motion for the electrons are as follows:

$$-\frac{K_1 Z'(\bar{r}_a - \sigma \bar{i})}{|\bar{r}_a - \sigma \bar{i}|^3} - \frac{K_1 Z'(\bar{r}_a + \sigma \bar{i})}{|\bar{r}_a + \sigma \bar{i}|^3} + \sum_b \frac{K_1(\bar{r}_a - \bar{r}_b)}{|\bar{r}_a - \bar{r}_b|^3} = m\frac{d^2\bar{r}_a}{dt^2} \quad (2.1)$$

where $a, b = 1, 2, \ldots, N_b$ and $a \neq b$. Here, $\bar{r}_a$ is the position vector of the e_a electron, which has its origin in the origin of the Cartesian coordinate axes, Z' is an order number which includes the effect of the 1s electrons. It is given by the relation $Z' = Z_A - 2s_{a,A1}$.

This system has three exact solutions, similar to the solutions of Eq. (1.72). There is one solution to each one of the following three cases: (a) for $N_b = 2$, which corresponds to single bond; (b) for $N_b = 4$, which corresponds to the double bond; (c) for $N_b = 6$, which corresponds to the triple bond.

2.2.1.1 Single Bond ($N_b = 2$)

The solution of the system in this case is:

$$\bar{r}_1 = -\bar{r}_2 = \bar{r} \quad \text{for} \quad \bar{r} \cdot \bar{k} = 0 \tag{2.2}$$

In virtue of Eq. (2.2), the solution of the system represented by Eq. (2.1) reduces to the solution of the following equation:

$$-\frac{K_1 Z'(\bar{r}_a - \sigma \bar{i})}{|\bar{r}_a - \sigma \bar{i}|^3} - \frac{K_1 Z'(\bar{r}_a + \sigma \bar{i})}{|\bar{r}_a + \sigma \bar{i}|^3} + \frac{K_1 \bar{r}_a}{4|\bar{r}_a|^3} = m \frac{d^2 \bar{r}_a}{dt^2} \tag{2.3}$$

which is valid for $a = 1$ or $a = 2$. This solution corresponds to symmetrical velocities. The symmetry of the velocities results from the total derivative of Eq. (2.2) with respect to time, which gives $\bar{v}_1 = -\bar{v}_2 = \bar{v}$. Equation (2.3) is equivalent to the following two scalar equations, in the plane xy:

$$-\frac{K_1 Z'(x - \sigma)}{[(x-\sigma)^2 + y^2]^{3/2}} - \frac{K_1 Z'(x + \sigma)}{[(x+\sigma)^2 + y^2]^{3/2}} + \frac{K_1 x}{4(x^2 + y^2)^{3/2}} = m \frac{d^2 x}{dt^2} \tag{2.4}$$

$$-\frac{K_1 Z' y}{[(x-\sigma)^2 + y^2]^{3/2}} - \frac{K_1 Z' y}{[(x+\sigma)^2 + y^2]^{3/2}} + \frac{K_1 y}{4(x^2 + y^2)^{3/2}} = m \frac{d^2 y}{dt^2} \tag{2.5}$$

The numerical solution of the above system of equations is given by Popa (1999b), and it leads to the curve C_1 that corresponds to electron e_1. The curve C_2, corresponding to electron e_2, is symmetrical to the curve C_1.

The atomic curves for the fundamental states of atoms, calculated in Chapter 1, are ellipses with eccentricities very close to unity. For this reason, in this chapter, we restrict ourselves to the case in which the molecular C_1 curve is composed of two atomic curves, denoted by C_{1A} and C_{1B}, which are ellipses with eccentricities very close to unity, as

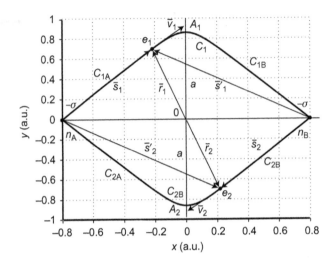

Figure 2.1 C_a *curves for single bond in homonuclear diatomic molecules.*

shown in Figure 2.1. At the scale of this figure, these ellipses are drawn as straight lines in the vicinities of the nuclei, in spite of the fact that the curve C_{1A} surrounds the nucleus n_A, while the curve C_{1B} surrounds the nucleus n_B. The curves C_1 and C_2 depend on three parameters: E_b, σ, and a, where E_b is the total energy of the particles that participate to the bond, and a is the maximum distance between the electron and the line on which the two nuclei are situated. In Figure 2.1, the electrons are situated in average positions, in agreement with Eq. (B.15) from Volume I.

2.2.1.2 Double Bond ($N_b = 4$)
In this case, the solution of the system represented by Eq. (2.1) is:

$$\bar{r}_1 = -\alpha\bar{i} + \beta\bar{j}, \quad \bar{r}_2 = -\alpha\bar{i} - \beta\bar{j}, \quad \bar{r}_3 = \alpha\bar{i} + \beta\bar{k},$$
$$\text{and} \quad \bar{r}_4 = \alpha\bar{i} - \beta\bar{k} \tag{2.6}$$

where α and β are real variables. It is easy to show that, in virtue of Eq. (2.6), the solution of the system represented by Eq. (2.1) reduces to the solution of the following equation, in which the unknown is $\bar{r}_a$:

$$-\frac{K_1 Z'(\bar{r}_a - \sigma\bar{i})}{|\bar{r}_a - \sigma\bar{i}|^3} - \frac{K_1 Z'(\bar{r}_a + \sigma\bar{i})}{|\bar{r}_a + \sigma\bar{i}|^3} + \frac{K_1[\bar{r}_a - (\bar{r}_a \cdot \bar{i})\bar{i}]}{4[\bar{r}_a^2 - (\bar{r}_a \cdot \bar{i})^2]^{3/2}}$$
$$+ \frac{K_1[\bar{r}_a + (\bar{r}_a \cdot \bar{i})\bar{i}]}{\sqrt{2}[\bar{r}_a^2 + (\bar{r}_a \cdot \bar{i})^2]^{3/2}} = m\frac{d^2\bar{r}_a}{dt^2} \tag{2.7}$$

This equation is valid for any bond electron, namely, for $a = 1, 2, 3, 4$. For example, Eq. (2.7) written for $\bar{r}_1$ is obtained from Eq. (2.1), in which we plug in the solutions of $\bar{r}_2$, $\bar{r}_3$, and $\bar{r}_4$ given by Eq. (2.6). Same procedure can be used to derive Eq. (2.7) for $a = 2, 3, 4$. The solution to Eq. (2.7) corresponds to symmetrical velocities. The symmetry of the velocities results from the total derivative of relations in Eq. (2.6) with respect to time.

Equation (2.7) leads again to a plane C_1 curve, for the electron e_1. It is equivalent to two scalar equations, similar to Eqs. (2.4) and (2.5). An identical numerical solution, as that given by Popa (1999b), leads again to a molecular curve, which is composed of two elliptic quasi-linear curves, when the minimum distance between electron and the nucleus is negligible. The curves for the other electrons are symmetrical, in agreement to Eq. (2.6). These curves, which result from Eq. (2.7), are shown in Figure 2.2. From this figure we see that, in the vicinities of the nuclei, C_1 and C_2 are two symmetrical curves, in the plane xy, similar to the valence curves in the case of the helium atom, or in the case of beryllium. The other curves, C_3 and C_4, are two symmetrical curves, with the same properties, situated in the plane xz.

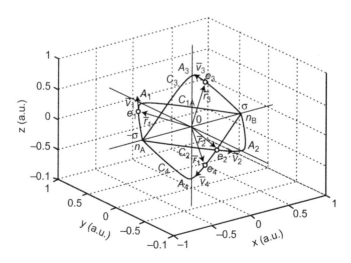

Figure 2.2 C_a *curves for double bond in homonuclear diatomic molecules.*

2.2.1.3 Triple Bond ($N_b = 6$)

In this case, the symmetrical solution of the system represented by Eq. (2.1) is:

$$\bar{r}_1 = -\alpha \bar{i} + \beta \bar{j} \tag{2.8}$$

$$\bar{r}_2 = -\alpha \bar{i} - \frac{1}{2}\beta \bar{j} + \frac{\sqrt{3}}{2}\beta \bar{k} \tag{2.9}$$

$$\bar{r}_3 = -\alpha \bar{i} - \frac{1}{2}\beta \bar{j} - \frac{\sqrt{3}}{2}\beta \bar{k} \tag{2.10}$$

$$\bar{r}_4 = \alpha \bar{i} - \beta \bar{j} \tag{2.11}$$

$$\bar{r}_5 = \alpha \bar{i} + \frac{1}{2}\beta \bar{j} - \frac{\sqrt{3}}{2}\beta \bar{k} \tag{2.12}$$

$$\bar{r}_6 = \alpha \bar{i} + \frac{1}{2}\beta \bar{j} + \frac{\sqrt{3}}{2}\beta \bar{k} \tag{2.13}$$

where α and β are real variables. It is easy to show that, as in the previous case, in virtue of Eqs. (2.8)–(2.13), the solution of the system represented by Eq. (2.1) reduces to the solution of the following equation in which the unknown is $\bar{r}_a$:

$$-\frac{K_1 Z'(\bar{r}_a - \sigma \bar{i})}{|\bar{r}_a - \sigma \bar{i}|^3} - \frac{K_1 Z'(\bar{r}_a + \sigma \bar{i})}{|\bar{r}_a + \sigma \bar{i}|^3} + \frac{K_1 \bar{r}_a}{4|\bar{r}_a|^3} + \frac{K_1[\bar{r}_a + 3(\bar{r}_a \cdot \bar{i})\bar{i}]}{[\bar{r}_a^2 + 3(\bar{r}_a \cdot \bar{i})^2]^{3/2}}$$
$$+ \frac{K_1[\bar{r}_a - (\bar{r}_a \cdot \bar{i})\bar{i}]}{\sqrt{3}[\bar{r}_a^2 - (\bar{r}_a \cdot \bar{i})^2]^{3/2}} = m\frac{d^2\bar{r}_a}{dt^2} \tag{2.14}$$

This equation is valid for any value of a. For example, by introducing the solutions for $\bar{r}_2$, $\bar{r}_3$, ... , $\bar{r}_6$, given by Eqs. (2.9)–(2.13), in Eq. (2.1) written for electron e_1, we obtain Eq. (2.14) for $a = 1$. In the same manner, we obtain the above equation for $a = 2, 3, ..., 6$. The solution to Eq. (2.14) corresponds again to symmetrical velocities.

Equation (2.14) leads to a plane C_1 curve for electron e_1. It is equivalent, as in the previous case, to two scalar equations, similar to Eqs. (2.4) and (2.5). An identical numerical solution, as that given by Popa (1999b), leads again to a molecular curve, which is composed of two elliptic quasi-linear curves. The curves for the other electrons are symmetrical, in agreement to Eqs. (2.8)–(2.13). These curves, which result from Eq. (2.14), are shown in Figure 2.3. From this figure we see that, in the vicinities of the nuclei, C_1, C_2, and C_3 are three symmetrical curves, similar to the curves of the valence electrons, in the

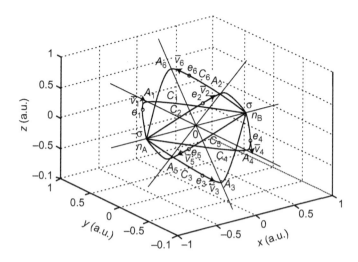

Figure 2.3 C_a *curves for triple bond in homonuclear diatomic molecules.*

case of the boron atom. The other curves, C_4, C_5, and C_6 are three symmetrical curves, with the same properties.

2.2.2 C_a Curves for Ionic and Covalent Bonds in Heteronuclear Molecules

We consider a heteronuclear molecule composed of two fixed different nuclei, denoted by n_A and n_B, two 1s electrons in the vicinity of the n_A nucleus, denoted by e_{A1} and e_{A2}, and two valence electrons which participate to the bond, denoted by e_1 and e_2, as shown in Figure 2.4. The Cartesian coordinates of the nuclei are $n_A(0,0,0)$ and $n_B(r_0,0,0)$ and their order numbers are, respectively, equal to Z_A and Z_B, where $Z_A > Z_B$. The equations of motion for electrons are as follows:

$$-\frac{K_1 Z'_A \bar{r}_a}{|\bar{r}_a|^3} - \frac{K_1 Z_B(\bar{r}_a - r_0\bar{i})}{|\bar{r}_a - r_0\bar{i}|^3} + \frac{K_1(\bar{r}_a - \bar{r}_b)}{|\bar{r}_a - \bar{r}_b|^3} = m\frac{d^2\bar{r}_a}{dt^2} \qquad (2.15)$$

where $a, b = 1, 2$ and $a \neq b$. Here, $\bar{r}_a$ is the position vector of the electron having the origin in the origin of the Cartesian coordinate axes, and Z'_A is an effective order number which includes the effect of the 1s electrons.

This system has the following symmetrical solution:

$$x_1 = x_2, \quad y_1 = -y_2, \quad \text{and} \quad z_1 = z_2 = 0 \qquad (2.16)$$

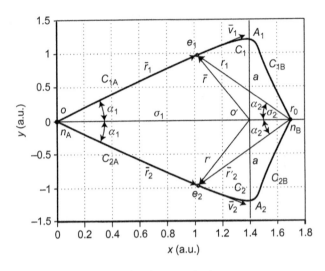

Figure 2.4 C_a *curves for ionic bond in heteronuclear diatomic molecules.*

The symmetry of the velocities results from the total derivative of these relations with respect to time.

An identical procedure, as that presented in the previous section, shows that, for this solution, the above system reduces to the following two scalar equations in the plane xy:

$$-\frac{K_1 Z'_A x_a}{(x_a^2 + y_a^2)^{3/2}} - \frac{K_1 Z_B (x_a - r_0)}{[(x_a - r_0)^2 + y_a^2]^{3/2}} = m\frac{d^2 x_a}{dt^2} \qquad (2.17)$$

$$-\frac{K_1 Z'_A y_a}{(x_a^2 + y_a^2)^{3/2}} - \frac{K_1 Z_B y_a}{[(x_a - r_0)^2 + y_a^2]^{3/2}} + \frac{K_1 y_a}{4|y_a|^3} = m\frac{d^2 y_a}{dt^2} \qquad (2.18)$$

which are valid for $a = 1, 2$.

The numerical solution of the above system of equations is almost identical to the solution of the system represented by Eqs. (2.4) and (2.5), which is given in Popa (1999b). The same numerical method presented in Popa (1999b) can be used to compute the curves C_1 and C_2 corresponding to the electrons e_1 and e_2, which are shown in Figure 2.4. In this figure, the coordinates of the points A_1 and A_2 are, respectively, $(\sigma_1, a, 0)$ and $(\sigma_1, -a, 0)$. We have also $\sigma_2 = r_0 - \sigma_1$. The difference between the curve C_1 resulted from Eqs. (2.4) and (2.5) and the curve C_1 resulted from Eqs. (2.17) and (2.18) is that the latter does not have an axis of symmetry parallel to the oy axis, as illustrated by the comparison between Figures 2.1 and 2.4, respectively. On the other hand, the curves C_1 and C_2 shown in Figure 2.4 are symmetric to each other with respect to the ox axis. Each of these curves is composed of two quasi-linear ellipses. The analysis of this figure shows the existence of two phases, the phase A, when the electrons are situated on the curves C_{1A} and C_{2A}, in the vicinity of the nucleus n_A, and the phase B, when the electrons are situated on the curves C_{1B} and C_{2B}, in the vicinity of the nucleus n_B. These curves correspond to the ionic bond because, due to the nonsymmetry, in the majority of the time, the electrons are in the vicinity of the nucleus whose order number is smaller.

A similar analysis, as that presented above, leads to the C_a curves corresponding to the covalent bond in heteronuclear diatomic molecule. In this case, the e_1 and e_2 electrons are situated in the vicinities of different nuclei, as shown in Figure 2.5. Unlike the ionic bond, in the covalent bond, the electrons move alternatively in the fields of different nuclei, and their charge is disposed in the vicinities of both nuclei.

In Section 2.3, we will show that the motion of the bond electrons is similar to that in a central field.

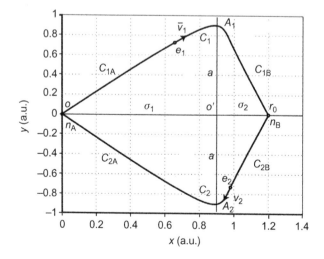

Figure 2.5 C_a *curves for covalent bond in heteronuclear diatomic molecules.*

2.3 CALCULATIONS OF GEOMETRIC PARAMETERS OF DIATOMIC MOLECULES

2.3.1 Geometric Parameters of Single Covalent Bond in Homonuclear Molecules

We consider a molecule composed of two fixed identical nuclei, denoted by n_A and n_B, four 1s electrons, and two valence electrons which participate to the bond, denoted by e_1 and e_2. The Cartesian coordinates of the nuclei are, respectively, $n_A(-\sigma, 0, 0)$ and $n_B(\sigma, 0, 0)$, and their order number is equal to Z_A. The C_1 and C_2 curves, which correspond to the bond electrons are situated in the plane xy, as shown in Figure 2.1. In Section 2.2.1, we have shown that each curve is composed of two quasi-linear ellipses, denoted by C_{aA} and C_{aB}. We assume that the kinetic energies of the electrons positioned at points $A_1(0, a, 0)$ and $A_2(0, -a, 0)$ are negligible, as compared to the total energy. The electrons are situated in average positions in Figure 2.1, and their coordinates are $e_1[-(1/4)\sigma, (3/4)a, 0]$ and $e_2[(1/4)\sigma, -(3/4)a, 0]$, in agreement with Eq. (B.15) from Volume I, for $r_M = (a^2 + \sigma^2)^{1/2}$. In this case, E_b has the significance of the total energy from which the energies of the 1s electrons are subtracted. The three quantities E_b, σ, and a are calculated with the aid of the following three relations.

2.3.1.1 The First Relation

This relation results from the central field approximation. It results in a similar manner as in the case of atoms, presented in Chapter 1: we have to write the expression of the energy of the electron e_1, denoted by E_1, and to consider the approximation of the central field. We have $E_1 = T_1 + U_{e_1 n_A}$, where T_1 is the kinetic energy of the electron e_1 and $U_{e_1 n_A}$ is the electrostatic energy of interaction between the electron e_1 and the nucleus n_A, which includes the screening effect of the 1s electrons. Taking into account the symmetry of the system (from where we have $T_1 = T_2$, $U_{e_1 n_A} = U_{e_2 n_B}$, $U_{e_1 n_B} = U_{e_2 n_A}$) and the expression of the energy E_b, we have:

$$E_1 = T_1 + U_{e_1 n_A} = \frac{E_b}{2} - U_0 \quad \text{with} \quad U_0 = U_{e_1 n_B} + \frac{1}{2}(U_{e_1 e_2} + U_{n_A n_B})$$

$$(2.19)$$

The significance of the above electrostatic terms is specified by their notations (e.g., $U_{e_1 n_B}$ is the potential energy of electrostatic interaction between the electron e_1 and nucleus n_B), and the screening effect of the

1s electrons is taken into account through the terms $U_{e_1 n_B}$ and $U_{n_A n_B}$. In these terms, which refer at the interaction at distance between electrons and nuclei, and, respectively, between nuclei, the screening coefficient due to the 1s electrons is equal to unity. The expressions of these terms are (Figure 2.1) $U_{e_1 n_B} = K_1 Z_{1M}/s'_1$ and $U_{n_A n_B} = K_1 Z_{nA}^2/(2\sigma)$, where the corresponding order numbers are:

$$Z_{1M} = Z_A - 2 \quad \text{and} \quad Z_{nA} = Z_A - 2 \tag{2.20}$$

Since U_0 is a relatively small quantity, we consider its averaged value, denoted by U_{0m}, calculated for the average positions of the electrons. In this case, the expression $E_1 = T_1 + U_{e_1 n_A} = E_b/2 - U_{0m}$ represents the constant energy of a hydrogenoid system, resulting that the electron e_1 moves in the central field of the nucleus in an averaged field of the other electron. We apply the quantization relation from Eq. (1.43) from Volume I, which is $\Delta_{C_{1A}} S_{01} = n_1 h$, for the curve C_{1A} and, in virtue of the relation (B.10) from Volume I, we have:

$$E_1 = \frac{E_b}{2} - U_{0m} = -\frac{Z_1^2 R_\infty}{n_1^2} \tag{2.21}$$

where

$$U_{0m} = -\frac{4K_1 Z_{1M}}{\sqrt{9a^2 + 25\sigma^2}} + \frac{K_1}{\sqrt{9a^2 + \sigma^2}} + \frac{K_1 Z_{nA}^2}{4\sigma} \tag{2.22}$$

and

$$Z_1 = Z_A - 2s_{1,A_1} \tag{2.23}$$

From Eqs. (2.21) and (2.22), we obtain:

$$E_b = -2\frac{R_\infty Z_1^2}{n_1^2} - \frac{8K_1 Z_{1M}}{\sqrt{9a^2 + 25\sigma^2}} + \frac{2K_1}{\sqrt{9a^2 + \sigma^2}} + \frac{K_1 Z_{nA}^2}{2\sigma} \tag{2.24}$$

2.3.1.2 The Second Relation

This relation results from the virial theorem. Taking into account that the curve C_1 is symmetrical, the virial theorem can be written as:

$$E_b = -\tilde{T} = -\frac{1}{\tau}\int_\tau mv_1^2 \, dt = -\frac{\Delta_{C_{1A}} S_{01}}{\tau} = -\frac{n_1 h}{\tau} \tag{2.25}$$

where $\tilde{T}$ is the average value of the total kinetic energy and τ is the period corresponding to the curve C_{1A}. Since $\tau = 2\int_0^{r_M} ds_1/v_1$ where $v_1 = 0$ and $T_1 = 0$ for $s_1 = r_M$, it results that the domain in the vicinity of the point A_1, which is relatively far from nucleus, has a big weight in the calculation of the integral. In this case, we have $U_{e_1 n_A} \cong -K_1 Z_{1M}/s_1$, and using the first relation from Eq. (2.19), we obtain $E_b/2 - U_0 = T_1 - K_1 Z_{1M}/s_1 = -K_1 Z_{1M}/r_M$, or $T_1 = K_1 Z_{1M}/s_1 - K_1 Z_{1M}/r_M$. The expression of τ becomes:

$$\tau = 2 \int_0^{r_M} \frac{ds_1}{\sqrt{2T_1/m}} = \sqrt{(2m)} \int_0^{r_M} \frac{ds_1}{\sqrt{K_1 Z_{1M}/s_1 - K_1 Z_{1M}/r_M}}$$
$$= \pi \sqrt{\frac{m}{2K_1 Z_{1M}}} r_M^{3/2} \qquad (2.26)$$

Introducing the expression of r_M in Eq. (2.26), Eq. (2.25) becomes:

$$E_b = -\frac{n_1 h \sqrt{2K_1 Z_{1M}}}{\pi \sqrt{m}(a^2 + \sigma^2)^{3/4}} \qquad (2.27)$$

2.3.1.3 The Third Relation
This relation results from the equation of the energy E_b, when the electrons are situated at maximum distance from the nuclei, respectively, in the points $A_1(0, a, 0)$ and $A_2(0, -a, 0)$, as follows:

$$E_b = 4U_{e_1 n_A} + U_{e_1 e_2} + U_{n_A n_B} = -\frac{4K_1 Z_{1M}}{\sqrt{a^2 + \sigma^2}} + \frac{K_1}{2a} + \frac{K_1 Z_{nA}^2}{2\sigma} \qquad (2.28)$$

2.3.2 Geometric Parameters of Double Covalent Bond in Homonuclear Molecules
2.3.2.1 The First Relation
The procedure is identical to that for single bonds. We need, however, to take into account that four electrons participate in this case to the bond. Their average positions are $e_1[-(1/4)\sigma, (3/4)a, 0]$, $e_2[-(1/4)\sigma, -(3/4)a, 0]$, $e_3[(1/4)\sigma, 0, (3/4)a]$, and $e_4[(1/4)\sigma, 0, -(3/4)a]$, as shown in Figure 2.2. In this case, taking into account the symmetry relations $T_1 = T_2 = T_3 = T_4$, $U_{e_1 n_A} = U_{e_2 n_A} = U_{e_3 n_B} = U_{e_4 n_B}$, $U_{e_1 n_B} = U_{e_2 n_B} = U_{e_3 n_A} = U_{e_4 n_A}$, and $U_{e_1 e_3} = U_{e_1 e_4} = U_{e_2 e_3} = U_{e_2 e_4}$, Eq. (2.19) becomes $E_1 = T_1 + U_{e_1 n_A} = E_b/4 - U_0$ with $U_0 = U_{e_1 n_B} + (1/4)(4U_{e_1 e_3} + U_{n_A n_B})$. A procedure identical to the one presented in Section 2.3.1, while taking into account the quantization relation, leads to the equation $E_1 = -R_\infty Z_1^2/n_1^2$, where:

$$Z_1 = Z_A - 2s_{1,A1} - s_{1,2} \quad \text{and} \quad s_{1,2} = s_{2,1} = \frac{\sqrt{a^2 + \sigma^2}}{4a} \qquad (2.29)$$

The screening coefficient $s_{1,2}$ is calculated with Eq. (1.59). We have also:

$$Z_{1M} = Z_A - 2 - s_{1,2} \qquad (2.30)$$

In virtue of the above relations, the second term of the second member of Eq. (2.24) corresponds to $U_{e_1 n_B} + U_{e_2 n_B} + U_{e_3 n_A} + U_{e_4 n_A} = 4U_{e_1 n_B}$ instead of $2U_{e_1 n_B}$ obtained in the case of the single bond. We can further write $4U_{e_1 n_B} = 4K_1(Z_B - 2)/\sqrt{(5\sigma/4)^2 + (3a/4)^2} = 16K_1(Z_{1M} + s_{1,2})/\sqrt{25\sigma^2 + 9a^2}$. The third term of the second member of Eq. (2.24) corresponds to the interactions between the bond electrons e_1, e_2 and the bond electrons e_3, e_4, and, because of symmetry, it corresponds to $4U_{e_1 e_3}$, instead of $U_{e_1 e_2}$ obtained in the case of single bond. With these modifications, the correspondent of Eq. (2.24), in the case of double bond, is:

$$E_b = -4\frac{R_\infty Z_1^2}{n_1^2} - \frac{16K_1(Z_{1M} + s_{1,2})}{\sqrt{9a^2 + 25\sigma^2}} + \frac{8K_1}{\sqrt{4.5a^2 + \sigma^2}} + \frac{K_1 Z_{nA}^2}{2\sigma} \qquad (2.31)$$

2.3.2.2 The Second Relation

Since there are four bond electrons, the virial relation must be modified, namely, instead of Eq. (2.25), we have $E_b = -2n_1 h/\tau$. Using an identical procedure as before, it follows that the second relation, that is, Eq. (2.27), becomes:

$$E_b = -\frac{2n_1 h\sqrt{2K_1 Z_{1M}}}{\pi\sqrt{m}(a^2 + \sigma^2)^{3/4}} \qquad (2.32)$$

2.3.2.3 The Third Relation

This relation results directly, taking into account that the first term from the second member represents the sum of the interactions between the bond electrons e_1, e_2, e_3, e_4 (situated, respectively, at points A_1, A_2, A_3, A_4) and nuclei, and the second term represents the sum of the interaction between all the bond electrons, corresponding to these positions. The third relation, Eq. (2.28) must be rewritten as follows:

$$E_b = -\frac{8K_1(Z_{1M} + s_{1,2})}{\sqrt{a^2 + \sigma^2}} + \frac{K_1}{a} + \frac{2\sqrt{2}K_1}{a} + \frac{K_1 Z_{nA}^2}{2\sigma} \qquad (2.33)$$

2.3.3 Geometric Parameters of Triple Covalent Bond in Homonuclear Molecules

The solution is similar to that for single bonds. We have to take into account that, in this case, six electrons participate to the bond. Their average positions are:

$$
e_1 \left[-\frac{1}{4}\sigma, \frac{3}{4}a, 0 \right], \quad
e_2 \left[-\frac{1}{4}\sigma, -\frac{3}{4}\cdot\frac{1}{2}a, \frac{3}{4}\cdot\frac{\sqrt{3}}{2}a \right],
$$

$$
e_3 \left[-\frac{1}{4}\sigma, -\frac{3}{4}\cdot\frac{1}{2}a, -\frac{3}{4}\cdot\frac{\sqrt{3}}{2}a \right], \quad
e_4 \left[\frac{1}{4}\sigma, -\frac{3}{4}a, 0 \right]
$$

$$
e_5 \left[\frac{1}{4}\sigma, \frac{3}{4}\cdot\frac{1}{2}a, -\frac{3}{4}\cdot\frac{\sqrt{3}}{2}a \right], \quad \text{and} \quad
e_6 \left[\frac{1}{4}\sigma, \frac{3}{4}\cdot\frac{1}{2}a, \frac{3}{4}\cdot\frac{\sqrt{3}}{2}a \right]
$$

as shown in Figure 2.3. The analysis in this case is analogous to the analysis of the single and double bonds. The difference from these two previous cases is that we have to take into account that six bond electrons participate to the triple bond instead of four (double bond) or two (single bond) electrons. It is easy to see that the three relations, in the case of the triple bond, become:

$$
E_b = -6\frac{R_\infty Z_1^2}{n_1^2} - \frac{24K_1(Z_{1M}+2s_{1,2})}{\sqrt{9a^2+25\sigma^2}} + \frac{6K_1}{\sqrt{9a^2+\sigma^2}} + \frac{24K_1}{\sqrt{9a^2+4\sigma^2}} + \frac{K_1 Z_{nA}^2}{2\sigma}
\tag{2.34}
$$

$$
E_b = -\frac{3n_1 h\sqrt{2K_1 Z_{1M}}}{\pi\sqrt{m}(a^2+\sigma^2)^{3/4}}
\tag{2.35}
$$

$$
E_b = -\frac{12K_1(Z_{1M}+2s_{1,2})}{\sqrt{a^2+\sigma^2}} + \frac{6K_1}{\sqrt{3}a} + \frac{6K_1}{a} + \frac{3K_1}{2a} + \frac{K_1 Z_{nA}^2}{2\sigma}
\tag{2.36}
$$

where

$$
Z_1 = Z_A - 2s_{1,A1} - 2s_{1,2}, \quad Z_{1M} = Z_A - 2 - 2s_{1,2}
\tag{2.37}
$$

and

$$
s_{1,2} = s_{2,1} = \frac{\sqrt{2}}{4\sqrt{1 - \frac{\sigma^2 - a^2/2}{\sigma^2 + a^2}}}
\tag{2.38}
$$

2.3.4 Geometric Parameters of Ionic Bonds in Heteronuclear Molecules

We consider a heteronuclear molecule composed of two fixed different nuclei, denoted by n_A and n_B, two 1s electrons in the vicinity of the n_A nucleus, denoted by e_{A1} and e_{A2}, and two valence electrons which participate to the bond, denoted by e_1 and e_2. In Section 2.2.2, we have shown that the curves C_1 and C_2 for this molecule are represented in Figure 2.4. These curves are situated in the plane xy, the Cartesian coordinates of the nuclei are $n_A(0,0,0)$ and $n_B(\sigma_1 + \sigma_2, 0, 0)$, and their order numbers are, respectively, equal to Z_A and Z_B, where $Z_A > Z_B$. The curve C_a, where $a = 1$ or $a = 2$, is composed of two quasi-linear ellipses, denoted by C_{aA} and C_{aB}, which surround, respectively, the nuclei n_A and n_B. We distinguish two phases: the phase A, when the electrons are situated on the curves C_{1A} and C_{2A}, in the vicinity of the nucleus n_A, and the phase B, when the electrons are situated on the curves C_{1B} and C_{2B}, in the vicinity of the nucleus n_B. The average positions of the electrons in the phase A are $e_1[(3/4)\sigma_1, (3/4)a, 0]$ and $e_2[(3/4)\sigma_1, -(3/4)a, 0]$, while in the phase B they are $e_1[\sigma_1 + (1/4)\sigma_2, (3/4)a, 0]$ and $e_2[\sigma_1 + (1/4)\sigma_2, -(3/4)a, 0]$. We note that the curves C_{1A} and C_{2A}, on one hand, and the curves C_{1B} and C_{2B}, on the other hand, are very similar to the curves C_1 and C_2 in the case of helium atom.

In the case of the heteronuclear diatomic molecules, the order number corresponding to the bond electron e_i is denoted by $Z_{i(A)}$ when e_i moves in the vicinity of the nucleus n_A and by $Z_{i(B)}$ when e_i moves in the vicinity of the nucleus n_B. The screening coefficients between the bond electrons e_1 and e_2, which move in the vicinity of the nucleus n_A, are denoted by $s_{1,2(A)}$ and $s_{2,1(A)}$. The screening coefficients between the electrons e_{Ai} and e_j when e_j moves in the vicinity of the nucleus n_A have the same significance, as they do in the case of the homonuclear molecules, and therefore, we keep the same notations for them, namely, $s_{Ai,j}$ and $s_{j,Ai}$.

The treatment of this system is almost identical to that presented in Section 2.3.1 for the single bond, with the difference that, due to asymmetry, the order numbers are different, for the two phases, as follows:

$$Z_{1(A)} = Z_A - 2s_{1,A1} - s_{1,2(A)} \tag{2.39}$$

$$Z_{1M(A)} = Z_A - 2 - s_{1,2(A)} \quad \text{and} \quad Z_{nA} = Z_A - 2 \tag{2.40}$$

$$Z_{1(B)} = Z_B - s_{1,2(B)} \tag{2.41}$$

$$Z_{1M(B)} = Z_B - s_{1,2(B)} \quad \text{and} \quad Z_{nB} = Z_B \tag{2.42}$$

where

$$s_{1,2(A)} = \frac{\sqrt{\sigma_1^2 + a^2}}{4a} \quad \text{and} \quad s_{1,2(B)} = \frac{\sqrt{\sigma_2^2 + a^2}}{4a} \tag{2.43}$$

Because of this difference between phases, the quantum numbers associated to the motion of the electron e_1 on curves C_{1A} and C_{1B}, denoted, respectively, by $n_{1(A)}$ and $n_{1(B)}$, and the corresponding periods, denoted by τ_A and τ_B, are different.

Due to the symmetry of the system, we have $Z_{1(A)} = Z_{2(A)}$, $Z_{1(B)} = Z_{2(B)}$, $Z_{1M(A)} = Z_{2M(A)}$, $Z_{1M(B)} = Z_{2M(B)}$, $n_{1(A)} = n_{2(A)}$, and $n_{1(B)} = n_{2(B)}$.

The kinetic energy of one electron, corresponding to the points A_1 or A_2 is denoted by T_m. Despite the fact that its value is very small, is it not neglected. The three types of equations from the previous sections are derived in a similar manner, as follows.

2.3.4.1 The First Type of Relation

These relations result from the central field approximation, using the quantization condition. The energy of the electron e_1, when the system is in the phase A, is:

$$E_1 = T_1 + U_{e_1 n_A} = \frac{E_b}{2} - U_{0A} \quad \text{with} \quad U_{0A} = U_{e_1 n_B} + \frac{1}{2} U_{n_A n_B} \tag{2.44}$$

Since the e_1 and e_2 electrons are situated in the vicinity of the nucleus n_A, their interaction energy is included in $U_{e_1 n_A}$ and $U_{e_2 n_A}$, by the terms $s_{1,2(A)}$ and $s_{1,2(B)}$.

The quantization relation, applied for the curve C_{1A}, is $\Delta_{C_{1A}} S_{01} = n_{1(A)} h$. The processing of these relations is identical to the processing of the relations (2.19), (2.20), and (2.21). Thus, we approximate the quantity U_{0A}, with its value, denoted by U_{0Am}, calculated when the electrons are in average positions. In this case, the motions of the electrons reduces to that in a central field. An identical procedure to that presented in Section 2.3.1 leads to the following relation:

$$E_b = -2\frac{R_\infty Z_{1(A)}^2}{n_{1(A)}^2} - \frac{8K_1[Z_{1M(B)} + s_{1,2(B)}]}{\sqrt{9a^2 + (\sigma_1 + 4\sigma_2)^2}} + \frac{K_1 Z_{nA} Z_{nB}}{\sigma_1 + \sigma_2} \tag{2.45}$$

The same analysis, made for the phase B, leads to the following relation:

$$E_b = -2\frac{R_\infty Z_{1(B)}^2}{n_{1(B)}^2} - \frac{8K_1[Z_{1M(A)} + s_{1,2(A)}]}{\sqrt{9a^2 + (\sigma_2 + 4\sigma_1)^2}} + \frac{K_1 Z_{nA} Z_{nB}}{\sigma_1 + \sigma_2} \tag{2.46}$$

2.3.4.2 The Second Type of Relation

These relations result from the virial theorem. We show that this theorem can be written separately for phases A and B. We take into account the notations from Figure 2.4 and write the following relation for the phase A: $d(m\bar{r} \cdot \bar{v}_1)/dt + d(m\bar{r}' \cdot \bar{v}_2)/dt = m\bar{r} \cdot d\bar{v}_1/dt + m\bar{r}' \cdot d\bar{v}_2/dt + mv_1^2 + mv_2^2$. The integration of this relation with respect to time, over the period τ_A, considering that $\bar{r} \cdot \bar{v}_1 \cong 0$ and $\bar{r}' \cdot \bar{v}_2 \cong 0$, respectively, at points A_1 and A_2, leads to the following relation:

$$2\tilde{T} = -\frac{1}{\tau_A}\int_{\tau_A} (\bar{r} \cdot \bar{F}_1 + \bar{r}' \cdot \bar{F}_2)dt \tag{2.47}$$

where $\bar{F}_1$ and $\bar{F}_2$ are the forces which act, respectively, on the electrons e_1 and e_2. With the aid of their expressions (Figure 2.4), and taking into account the symmetry of the system and the fact that $\bar{r} = -\sigma_1\bar{i} + \bar{r}_1 = \sigma_2\bar{i} + \bar{r}_1'$ and $\bar{r}' = -\sigma_1\bar{i} + \bar{r}_2 = \sigma_2\bar{i} + \bar{r}_2'$, we obtain:

$$\bar{r} \cdot \bar{F}_1 + \bar{r}' \cdot \bar{F}_2 = \left[-\frac{K_1 Z_{1(A)}}{r_1^3}\bar{r}_1 - \frac{K_1 Z_{1M(B)}}{r_1'^3}\bar{r}_1' + \frac{K_1}{|\bar{r}_1 - \bar{r}_2|^3}(\bar{r}_1 - \bar{r}_2)\right]\bar{r}$$

$$+ \left[-\frac{K_1 Z_{1(A)}}{r_2^3}\bar{r}_2 - \frac{K_1 Z_{1M(B)}}{r_2'^3}\bar{r}_2' + \frac{K_1}{|\bar{r}_1 - \bar{r}_2|^3}(\bar{r}_2 - \bar{r}_1)\right]\bar{r}'$$

$$= 2\sigma_1 \frac{K_1 Z_{1(A)}}{r_1^2}\cos\alpha_1 + 2\sigma_2 \frac{K_1 Z_{1M(B)}}{r_1'^2}\cos\alpha_2 + U - \frac{K_1 Z_{nA} Z_{nB}}{\sigma_1 + \sigma_2} \tag{2.48}$$

where $U = -2K_1 Z_{1(A)}/r_1 - 2K_1 Z_{1M(B)}/r_1' + K_1/|\bar{r}_1 - \bar{r}_2| + K_1 Z_{nA} Z_{nB}/(\sigma_1 + \sigma_2)$. This is the potential energy of the system.

Introducing Eq. (2.48) in Eq. (2.47) and supposing that the average forces which act on the two nuclei are zero, we obtain $2\tilde{T} = -\tilde{U}$. Since $E_b = \tilde{U} + \tilde{T}$, we have:

$$E_b = -\tilde{T} = -\frac{1}{\tau_A}\int_{\tau_A} mv_1^2 \, dt = -\frac{\Delta_{C_{1A}}S_{01}}{\tau_A} = -\frac{n_{1(A)}h}{\tau_A} \tag{2.49}$$

In this case, $\tau_A = 2\int_0^{r_{M(A)}} dr_1/v_1 = 2\int_0^{r_{M(A)}} dr_1/\sqrt{2T_1/m}$, where $T_1 = T_m$ for $r_1 = r_{M(A)} = \sqrt{\sigma_1^2 + a^2}$. Since T_m is a very small quantity, we consider again that the domain in the vicinity of the point A_1 has a big weight in the calculation of the integral. We, thus, have $U_{e_1 n_A} \cong -K_1 Z_{1M(A)}/r_1$. The calculation of τ_A is performed similarly as in Eq. (2.26). Using Eq. (2.44), it follows that $E_b/2 - U_{0A} = T_1 - K_1 Z_{1M(A)}/r_1 = T_m - K_1 Z_{1M(A)}/r_{M(A)}$, or $T_1 = K_1 Z_{1M(A)}/r_1 - K_1 Z_{1M(A)}/r_{M(A)} + T_m$ and, instead of Eq. (2.26), we write:

$$\tau_A = \sqrt{2m}\int_0^{r_{M(A)}} \frac{dr_1}{\sqrt{\frac{K_1 Z_{1M(A)}}{r_1} - \frac{K_1 Z_{1M(A)}}{r_{M(A)}} + T_m}} = \pi\sqrt{\frac{m}{2K_1 Z_{1M(A)}}} r_{M(A)}^{3/2} t_{cA} \tag{2.50}$$

where t_{cA} is a correction term, given by the relation:

$$t_{cA} = \frac{2}{\pi a_{cA}^{3/2}}\left(\arcsin\sqrt{a_{cA}} - \sqrt{a_{cA}}\sqrt{1 - a_{cA}}\right) \tag{2.51}$$

with

$$a_{cA} = 1 - \frac{T_m\sqrt{\sigma_1^2 + a^2}}{K_1 Z_{1M(A)}} \tag{2.52}$$

Since $r_{M(A)} = \sqrt{\sigma_1^2 + a^2}$, from Eqs. (2.49) and (2.50), we obtain:

$$E_b = -\frac{n_{1(A)}h\sqrt{2K_1 Z_{1M(A)}}}{\pi\sqrt{m}(\sigma_1^2 + a^2)^{3/4} t_{cA}} \tag{2.53}$$

An identical analysis, corresponding to phase B, leads to the following relations:

$$E_b = -\frac{n_{1(B)}h\sqrt{2K_1 Z_{1M(B)}}}{\pi\sqrt{m}(\sigma_2^2 + a^2)^{3/4} t_{cB}} \tag{2.54}$$

where

$$t_{cB} = \frac{2}{\pi a_{cB}^{3/2}} (\arcsin \sqrt{a_{cB}} - \sqrt{a_{cB}}\sqrt{1 - a_{cB}}) \tag{2.55}$$

and

$$a_{cB} = 1 - \frac{T_m \sqrt{\sigma_2^2 + a^2}}{K_1 Z_{1M(B)}} \tag{2.56}$$

2.3.4.3 The Third Type of Relation

This relation, which is the expression of the energy E_b, when the electrons are situated, respectively, at points $A_1(0, a, 0)$ and $A_2(0, -a, 0)$, results directly:

$$E_b = -\frac{2K_1[Z_{1M(A)} + s_{1,2(A)}]}{\sqrt{\sigma_1^2 + a^2}} - \frac{2K_1[Z_{1M(B)} + s_{1,2(B)}]}{\sqrt{\sigma_2^2 + a^2}} + \frac{K_1}{2a} + \frac{K_1 Z_{nA} Z_{nB}}{\sigma_1 + \sigma_2} + 2T_m \tag{2.57}$$

2.3.5 Geometric Parameters of Covalent Bonds in Heteronuclear Molecules

We consider the same heteronuclear covalent diatomic molecule as that presented in Section 2.2.2, having $n_A > n_B$, when the e_1 and e_2 electrons are situated in the vicinities of different nuclei, as shown in Figure 2.5. The average positions of the electrons are $e_1[(3/4)\sigma_1, (3/4)a, 0]$ and $e_2[\sigma_1 + (1/4)\sigma_2, -(3/4)a, 0]$. The curves C_1 and C_2 are asymmetrical, similar to those shown in Figure 2.4. The treatment of this system is almost identical to that from Section 2.3.1 for single bond, with the difference that the order numbers are different. For the positions of the electrons shown in Figure 2.5, we have:

$$Z_{1(A)} = Z_A - 2s_{1,A1}, \quad Z_{1M(A)} = Z_A - 2, \quad \text{and} \quad Z_{nA} = Z_A - 2 \tag{2.58}$$

$$Z_{2(B)} = Z_B, \quad Z_{2M(B)} = Z_B, \quad \text{and} \quad Z_{nB} = Z_B \tag{2.59}$$

The quantum numbers associated to the electrons e_1 and e_2, for the positions shown in Figure 2.5, are respectively, $n_{1(A)}$ and $n_{2(B)}$. Taking into account the symmetry of the curves C_1 and C_2, we have $Z_{1(A)} = Z_{2(A)}$, $Z_{1(B)} = Z_{2(B)}$, $Z_{1M(A)} = Z_{2M(A)}$, $Z_{1M(B)} = Z_{2M(B)}$, $n_{1(A)} = n_{2(A)}$, and $n_{1(B)} = n_{2(B)}$.

Both electrons move alternatively in the fields of the two nuclei, on the curves C_{1A}, C_{1B} and C_{2A}, C_{2B}. We assume that the kinetic energies of the bond electrons at the points A_1 and A_2 are neglected, as in Section 2.3.1, and that the motion of the electrons is periodical. The periodicity leads to the following relation:

$$\tau_A = \tau_B \tag{2.60}$$

In the case of the ionic bond, the electrons move together on helium type trajectories in the field of the same nucleus, and due to the non-symmetry of the C_a trajectory, in the majority of the time, the electrons are in the vicinity of the nucleus whose order number is smaller. Unlike the ionic bond, in the present case, the electrons move alternatively in the fields of different nuclei, the two periods of motion, τ_A and τ_B, are equal and their charge is disposed in the vicinities of both nuclei, from here resulting the covalent character of the bond.

The three types of equations result similarly as in Section 2.3.1, which are described as follows.

2.3.5.1 The First Type of Relation

This relation results from the central field approximation, using the quantization condition. In this case, the expression of the energy E_b can be written as:

$$E_b = E_1 + E_2 + U_{01} + U_{02} \tag{2.61}$$

with

$$E_1 = T_1 + U_{e_1 n_A} \quad \text{and} \quad E_2 = T_2 + U_{e_2 n_B} \tag{2.62}$$

where

$$U_{01} = U_{e_1 n_B} + \frac{1}{2}(U_{e_1 e_2} + U_{n_A n_B}) \tag{2.63}$$

and

$$U_{02} = U_{e_2 n_A} + \frac{1}{2}(U_{e_1 e_2} + U_{n_A n_B}) \tag{2.64}$$

We consider that the motions of the e_1 and e_2 electrons are approximated by motions in the central fields of the nuclei n_A and n_B. In this case, the quantities U_{01} and U_{02} are calculated for the average positions of the electrons, resulting that E_1 and E_2 are constants. We apply quantization relations $\Delta_{C_{1A}} S_{01} = n_{1(A)} h$ and $\Delta_{C_{2B}} S_{02} = n_{2(B)} h$, respectively, for the curves C_{1A} and C_{2B}. We obtain $E_1 = -R_\infty Z_{1(A)}^2 / n_{1(A)}^2$ and $E_2 = -R_\infty Z_{2(B)}^2 / n_{2(B)}^2$. Taking into account Eqs. (2.61)–(2.64), we obtain:

$$
E_b = -\frac{R_\infty Z_{1(A)}^2}{n_{1(A)}^2} - \frac{R_\infty Z_{2(B)}^2}{n_{2(B)}^2} - \frac{4K_1 Z_{1M(B)}}{\sqrt{9a^2 + (\sigma_1 + 4\sigma_2)^2}}
$$

$$
- \frac{4K_1 Z_{1M(A)}}{\sqrt{9a^2 + (\sigma_2 + 4\sigma_1)^2}} + \frac{4K_1}{\sqrt{36a^2 + (\sigma_1 + \sigma_2)^2}} + \frac{K_1 Z_{nA} Z_{nB}}{\sigma_1 + \sigma_2}
$$

$$(2.65)$$

2.3.5.2 The Second Type of Relation

The virial theorem, written for the system represented by Figure 2.5, and taking into account the symmetry of the curves C_1 and C_2, is:

$$
E_b = -\frac{1}{\tau_A} \int_{\tau_A} \frac{mv_1^2}{2}\, dt - \frac{1}{\tau_B} \int_{\tau_B} \frac{mv_2^2}{2}\, dt = -\frac{\Delta_{C_{1A}} S_{01}}{2\tau_A} - \frac{\Delta_{C_{2B}} S_{02}}{2\tau_B}
$$

$$
= -\frac{n_{1(A)} h}{2\tau_A} - \frac{n_{2(B)} h}{2\tau_B}
$$

$$(2.66)$$

which, in virtue of Eq. (2.60), can be written as:

$$
2\tau_A E_b = 2\tau_B E_b = -[n_{1(A)} + n_{1(B)}] h \tag{2.67}
$$

A calculation similar to that presented in Section 2.3.1, which takes into account Eq. (2.62), leads to the following relation:

$$
\tau_A = \tau_B = \pi \sqrt{\frac{m}{2K_1 Z_{1M(A)}}} (\sigma_1^2 + a^2)^{3/4}
$$

$$
= \pi \sqrt{\frac{m}{2K_1 Z_{1M(B)}}} (\sigma_2^2 + a^2)^{3/4} = -\frac{[n_{1(A)} + n_{1(B)}] h}{2E_b}
$$

$$(2.68)$$

2.3.5.3 The Third Type of Relation

This relation, which is the expression of the energy E_b, when the electrons are situated, respectively, at points $A_1(0, a, 0)$ and $A_2(0, -a, 0)$, results directly:

$$
E_b = -\frac{2K_1 Z_{1M(A)}}{\sqrt{\sigma_1^2 + a^2}} - \frac{2K_1 Z_{1M(B)}}{\sqrt{\sigma_2^2 + a^2}} + \frac{K_1}{2a} + \frac{K_1 Z_{nA} Z_{nB}}{\sigma_1 + \sigma_2} \tag{2.69}
$$

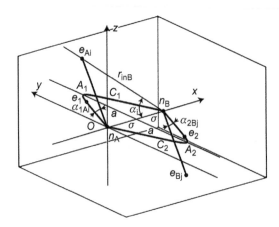

Figure 2.6 General structure of a homonuclear diatomic molecule.

2.4 ANALYTICAL METHOD USED TO CALCULATE THE ENERGETIC VALUES OF DIATOMIC MOLECULES

In this section, we present a general analytical method for the calculation of the energetic values and geometric parameters of diatomic molecules.

The structure of the molecule is illustrated in Figure 2.6. Despite the fact that in this figure we consider the case of a homonuclear molecule with a single bond, the method presented in this section is valid for all types of bonds. The molecule shown in Figure 2.6 is composed of two nuclei denoted by n_A and n_B and having $Z_A = Z_B$, four 1s electrons, denoted by e_{A1}, e_{A2}, e_{B1}, and e_{B2}, two electrons which participate to the bond, denoted by e_1 and e_2, N_A electrons which move in the field of the nucleus n_A and do not participate to the bond, denoted by e_{Ai}, and N_B electrons which move in the field of the nucleus n_B and do not participate to the bond, denoted by e_{Bj}. The electrons e_{Ai} and e_{Bj} are shown in average positions in Figure 2.6. We suppose that the molecule is symmetrical, and has $N_A = N_B$ and $i, j = 3, 4, \ldots, N_A$.

The calculation of the energetic values and geometric parameters of the diatomic molecule is performed in two stages, as follows.

2.4.1.1 The First Stage
We calculate the values of E_b, σ, and a with the aid of Eqs. (2.24), (2.27), and (2.28). We use normalized quantities. The energies are

normalized to R_∞ and the distances to $2a_0$, as follows: $\underline{E}_b = E_b/R_\infty$, $\underline{\sigma} = \sigma/2a_0$, and $\underline{a} = a/2a_0$. Taking into account these relations, together with Eq. (B.10) from Volume I, which is $R_\infty = (mK_1^2)/(2\hbar^2) = K_1/(2a_0)$, Eqs. (2.24), (2.27), and (2.28) become:

$$\underline{E}_b = -2\frac{Z_1^2}{n_1^2} - \frac{8Z_{1M}}{\sqrt{9\underline{a}^2 + 25\underline{\sigma}^2}} + \frac{2}{\sqrt{9\underline{a}^2 + \underline{\sigma}^2}} + \frac{Z_{nA}^2}{2\underline{\sigma}} \tag{2.70}$$

$$\underline{E}_b = -\frac{2n_1\sqrt{Z_{1M}}}{(\underline{a}^2 + \underline{\sigma}^2)^{3/4}} \tag{2.71}$$

$$\underline{E}_b = -\frac{4Z_{1M}}{\sqrt{\underline{a}^2 + \underline{\sigma}^2}} + \frac{1}{2\underline{a}} + \frac{Z_{nA}^2}{2\underline{\sigma}} \tag{2.72}$$

This is a system of three equations with three unknowns: $\underline{E}_b$, $\underline{\sigma}$, and $\underline{a}$. In this case, taking into account the structure of the molecule, shown in Figure 2.6, the expression of Z_1 must contain the screening coefficients due to the e_{Ai} electrons, and it can be written as:

$$Z_1 = Z_A - 2s_{1,A1} - \sum_{i=3}^{N_A} s_{1,Ai} \tag{2.73}$$

where $s_{1,A1}$ is approximated very well by s_{31e}, calculated for atoms, while $s_{1,Ai}$ is calculated with Eq. (1.59). The expression of Z_{1M} is modified, as follows:

$$Z_{1M} = Z_A - 2 - \sum_{i=3}^{N_A} s_{1,Ai} \tag{2.74}$$

Also, the expression of Z_{nA} must be modified, in order to take into account the effect of the screening due to the electrons e_{Ai}. We suppose that the electron e_{Ai} has the same effect on nucleus n_B, as a charge $-es_{n_B,Ai}$ placed on nucleus n_A. The forces in the ox direction acting on a unit charge placed on the nucleus n_B are equal in the two cases. With the notations from Figure 2.6, we have:

$$\frac{K_1 Z_{nA}}{(2\sigma)^2} = \frac{K_1(Z - 2 - s_{n_B,Ai})}{(2\sigma)^2} = \frac{K_1(Z - 2)}{(2\sigma)^2} - \frac{K_1 \cos \alpha_i}{r_{inB}^2} \tag{2.75}$$

Taking into account the effect of all the electrons e_{Ai}, we have:

$$Z_{nA} = Z_A - 2 - \sum_{i=3}^{N_A} \frac{(2\sigma)^2}{r_{inB}^2} \cos \alpha_i \qquad (2.76)$$

2.4.1.2 The Second Stage

In this stage, we calculate directly the normalized expression of the total energy, corresponding to the average positions of the electrons. We assume that the average positions of the electrons which do not participate to the bond are disposed in a configuration having maximum symmetry. The average normalized distance between the e_{Ai} electron and the nucleus is calculated with the aid of the relations (B.14) and (B.15) from Volume I.

Taking into account the symmetry of the system, where $\underline{E}_1 = \underline{E}_2$, $\underline{E}_{Ai} = \underline{E}_{Bi}$, $\underline{E}_{A1} = \underline{E}_{A2} = \underline{E}_{B1} = \underline{E}_{B2}$, $\underline{U}_{e_1 n_B} = \underline{U}_{e_2 n_A}$, $\underline{U}_{e_{Ai} n_B} = \underline{U}_{e_{Bi} n_A}$, and $\underline{U}_{e_{Ai} e_2} = \underline{U}_{e_{Bi} e_1}$, the normalized expression of the total energy can be written as follows:

$$\underline{E} = 2\underline{E}_1 + 2\sum_{i=3}^{N_A}\underline{E}_{Ai} + 4\underline{E}_{A1} + \underline{U}_{e_1 e_2} + 2\underline{U}_{e_1 n_B} + 2\sum_{i=3}^{N_A}\underline{U}_{e_{Ai} e_2}$$

$$+ 2\sum_{i=3}^{N_A}\underline{U}_{e_{Ai} n_B} + \sum_{i=3}^{N_A}\sum_{j=3}^{N_B}\underline{U}_{e_{Ai} e_{Bj}} + \underline{U}_{n_A n_B} + 2\underline{E}_{mls} = -2\frac{Z_1^2}{n_1^2} - 2\sum_{i=3}^{N_A}\frac{Z_{Ai}^2}{n_{Ai}^2}$$

$$- 4\frac{Z_{A1}^2}{n_{A1}^2} + \frac{1}{|\bar{r}_{(e_1)} - \bar{r}_{(e_2)}|} - \frac{2(Z_A - 2)}{|\bar{r}_{(e_1)} - \bar{r}_{(n_B)}|} + 2\sum_{i=3}^{N_A}\frac{1}{|\bar{r}_{(e_{Ai})} - \bar{r}_{(e_2)}|}$$

$$- 2\sum_{i=3}^{N_A}\frac{Z_A - 2}{|\bar{r}_{(e_{Ai})} - \bar{r}_{(n_B)}|} + \sum_{i=3}^{N_A}\sum_{j=3}^{N_B}\frac{1}{|\bar{r}_{(e_{Ai})} - \bar{r}_{(e_{Bj})}|} + \frac{(Z_A - 2)^2}{|\bar{r}_{(n_A)} - \bar{r}_{(n_B)}|} + 2\frac{Z_{A1}^{3/2}}{8n_{A1}^3}$$

$$(2.77)$$

where $\bar{r}_{(e_1)}$ and $\bar{r}_{(n_A)}$ are, respectively, the average normalized position vectors of the electron e_1 and nucleus n_A, $\underline{U}_{e_1 e_2}$ is the normalized potential energy of electrostatic interaction between the electron e_1 and e_2, and so on. The expressions of the order numbers are Z_1 given by Eq. (2.73) and:

$$Z_{A1} = Z_A - s_{A1,A2} - s_{A1,1} - \sum_{i=3}^{N_A} s_{A1,Ai} \qquad (2.78)$$

$$Z_{Ai} = Z_A - 2s_{Ai,A1} - s_{Ai,1} - \sum_{j=3; i \neq j}^{N_A} s_{Ai,Aj} \quad \text{for } i \geq 3 \qquad (2.79)$$

where $s_{Ai,A1}$ are, with good approximation, equal to s_{31e}, calculated in the case of atoms, while $s_{A1,1}$ and $s_{A1,Ai}$ are equal to s_{13e}. The coefficients $s_{Ai,1}$, $s_{Ai,Aj}$, and $s_{A1,A2}$ are calculated with Eq. (1.59).

The first, second, and third terms of the second member in Eq. (2.77) are the total normalized energies of the electrons which move in the field of the same nucleus. Their significance is identical to the significance of electron energies in an atomic system (see Chapter 1), taking into account that new indexes are introduced in the case of molecular systems, as described in Section 2.1. We recall that the components of the energies, denoted by $\underline{E}_1$, $\underline{E}_{Ai}$, or $\underline{E}_{A1}$, are given by Eqs. (1.128) and (1.129). The other terms represent the normalized electrostatic interaction energies, corresponding to interactions at distance between electrons that do not move in the field of the same nucleus, between electrons and nuclei which are situated at distance from them, and between nuclei having order numbers which include the screening effects of the 1s. The last term is given by Eq. (1.10) and represents the normalized correction energies due to the spin magnetic interaction of the 1s electrons. An analysis of Eq. (2.77) shows that all the components of the interaction energies are taken into account.

The normalized experimental values of the total energy are obtained with the aid of the relation:

$$\underline{E}_{\exp} = -2\underline{S}_{EiA} - \underline{D}_0^0 \tag{2.80}$$

where S_{EiA} and D_0^0 are, respectively, the sum of the ionization energies of the atom A and the observed dissociation energy. The values of S_{EiA} and D_0^0 are taken, respectively, from Lide (2003) and Huber and Hertzberg (1979). In Table 2.1, we compare our theoretical results with similar results reported in literature (CCCBDB), obtained with the aid of the Hartree–Fock method and with the experimental values. The calculations are made with the aid of Mathematica 7, the absolute error being of the order 10^{-16}. The calculation programs are given in Appendix B.

In virtue of the hypothesis (h5) from Chapter 1 of Volume I, the kinetic energy of the nuclei is zero. It follows that the distance between nuclei, denoted by r_0 and calculated with the relation:

$$r_0 = 2\sigma \tag{2.81}$$

Table 2.1 Normalized Values of the Total Energies $\underline{E}$, Calculated for Homonuclear and Heteronuclear Diatomic Molecules

Molecule	$\underline{E}$	$\underline{E}_{HF}$	$\underline{E}_{exp}$
Li_2	-30.0919	-29.73625	-29.9896
		-29.736246	
Be_2	-58.7676	-58.486596	-58.8396
		-58.48645	
B_2	-98.3291	-98.182662	-98.8574
		-98.18259	
C_2	-151.148	-150.804028	-151.883
		-150.803916	
LiH	-16.1821	-15.967824	-16.1348
		-15.96775	
BeH	-30.4238	-30.300286	-30.4872
		-30.300248	
BH	-50.4587	-50.256246	-50.5691
		-50.256092	
CH	-76.6349	-76.561468	-76.968
		-76.560994	

For comparison are given the best two values of the normalized values of the total energies from CCCBDB, denoted by $\underline{E}_{HF}$, which are calculated with the aid of the Hartree−Fock method and the corresponding experimental values $\underline{E}_{exp}$. All the values are given in Rydbergs.

corresponds to values of the nuclei velocities of zero. On the other hand, the experimental value of the distance between nuclei, denoted by r_e, corresponds to the minimum of the Morse curve (Hertzberg, 1950). The theory of the harmonic oscillation of the nuclei shows that this minimum corresponds to maximum velocities of the nuclei. It follows that r_0 is slightly different from r_e because the first corresponds to zero velocities of the nuclei, while the last corresponds to the maximum velocities of the nuclei.

The values of r_0, calculated from Eq. (2.81), and the experimental value of the distance between nuclei, denoted by r_e and taken from CCCBDB, are given in angstroms in Table 2.2. Table 2.2 also gives the range of values for the distances between nuclei, denoted by r_{eHF}, calculated with the Hartree−Fock method and taken from CCCBDB.

In Section 2.5, we will present details of the calculations.

Table 2.2 Theoretical Values of r_0 Compared with the Experimental Values of r_e (CCCBDB), and Also with the Range of Values for r_{eHF} Calculated Using Hartree–Fock Method (CCCBDB)

Molecule	r_0	r_e	r_{eHF}
Li_2	3.10043	2.673	2.696–2.816
Be_2	2.212	2.460	1.780–2.049
B_2	1.77241	1.590	1.341–1.636
C_2	1.35694	1.243	1.240–1.308
LiH	1.84635	1.5957	1.511–1.636
BeH	1.4389	1.3426	1.301–1.356
BH	1.21642	1.2324	1.213–1.232
CH	1.07572	1.1199	1.105–1.119

All the values are given in angstroms.

Table 2.3 Data for Li_2 Molecule

Particle Coordinates; Quantum Numbers. Calculated and Experimental Values	Screening Coefficients; Order Numbers
$n_A(-\underline{a},0,0)$; $n_B(\underline{a},0,0)$	$s_{1,A1} = 0.854942$
$e_1\left(-\frac{1}{4}\underline{a},\frac{3}{4}\underline{a},0\right)$; $e_2\left(-\frac{1}{4}\underline{a},-\frac{3}{4}\underline{a},0\right)$	$s_{A1,1} = 0.0013792$
$n_1 = 2$; $n_{A1} = n_{A2} = 1$	$s_{A1,A2} = 0.25$
$\underline{E}_b = -1.0115$; $\underline{a} = 1.46474$	$Z_1 = Z_2 = Z_A - 2s_{1,A1}$
$\underline{a} = 2.02683$	$Z_{1M} = Z_{2M} = Z_A - 2$
$\underline{S}_{EiA} = 14.9563$ (Lide, 2003)	$Z_{A1} = Z_{A2} = Z_A - s_{A1,A2} - s_{A1,1}$
$\underline{D}_0^0 = 0.0768815$ (Huber and Hertzberg, 1979)	$Z_{nA} = Z_{nB} = Z_A - 2$

2.5 TYPICAL APPLICATIONS

2.5.1 The Li_2 Molecule

This molecule contains only valence electrons that participate to the bond. Its structure is shown in Figure 2.1. We follow the algorithm presented Section 2.4 and calculate the screening coefficients, order numbers, and then solve the system (2.70)–(2.72). All these values are given in Table 2.3. The screening coefficients between the 1s electrons and the e_1 electron, namely, $s_{1,A1}$ and $s_{A1,1}$ are identical, respectively, to s_{31e} and s_{13e}, which have been calculated in Section 1.4.1 for the $1s^2 2s$ state of lithium atom (see Table 1.4). The coefficient $s_{A1,A2}$ is calculated with the aid of Eq. (1.59). Since 2s valence electrons participate to the bond, we have $n_1 = 2$. Also, we have $n_{A1} = 1$ because this number corresponds to 1s electrons. We solve the system (2.70)–(2.72) in order to obtain $\underline{E}_b$, $\underline{a}$, and $\underline{a}$.

Since in this case $\underline{E}_b$ has the significance of the total energy from which the energies of the 1s electrons are subtracted, we have:

$$\underline{E} = \underline{E}_b + 4\underline{E}_{A1} + 2\underline{E}_{m1s} = \underline{E}_b - 4\frac{Z_{A1}^2}{n_{A1}^2} + \frac{Z_{A1}^{3/2}}{8n_{A1}^3} \qquad (2.82)$$

The normalized value of the total energy $\underline{E}$, obtained from Eq. (2.82), together with the experimental value, denoted by $\underline{E}_{exp}$, are presented in Table 2.1. We also present in this table the best two values of the normalized total energies, calculated with the aid of the Hartree–Fock method, denoted by $\underline{E}_{HF}$, which are taken from CCCBDB.

The experimental values of the total energy are obtained with the aid of Eq. (2.80), where S_{EiA} and D_0^0 are taken, respectively, from Lide (2003) and Huber and Hertzberg (1979). The values r_0 and r_e are given in Table 2.2. The calculations are made with a Mathematica 7 program, which is given in Section B.2.1.

2.5.2 The Be$_2$ Molecule

The structure of this molecule is shown in Figure 2.7A. This molecule contains two electrons, denoted by e_{A3} and e_{B3}, which do not participate to the bond. The geometrical structure of the molecule results from the system (2.70)–(2.72), where the expressions of Z_1, Z_{1M}, and Z_{nA} are given in Table 2.4. The screening coefficients between the 1s electrons and the e_1, e_2, e_{A3}, and e_{B3} electrons, namely, $s_{1,A1}$, $s_{A1,1}$, $s_{A3,A1}$, $s_{A1,A3}$, and so on, are identical, respectively, to s_{31e} and s_{13e}, which have been calculated in Chapter 1 for the $1s^2 2s^2$ state of

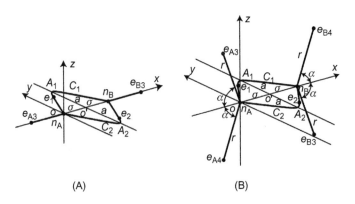

(A)   (B)

Figure 2.7 Structures of (A) Be$_2$ and (B) B$_2$ molecules.

Table 2.4 Data for Be$_2$ Molecule

Particle Coordinates; Quantum Numbers. Calculated and Experimental Values	Screening Coefficients; Order Numbers
$n_A(0,0,0)$; $n_B(2\underline{\sigma},0,0)$	$s_{1,A1} = s_{A3,A1} = 0.83882$
$e_1\left(\frac{3}{4}\underline{\sigma},\frac{3}{4}\underline{a},0\right)$; $e_1\left(\frac{5}{4}\underline{\sigma}, -\frac{3}{4}\underline{a},0\right)$	$s_{A1,1} = s_{A1,A3} = 0.00221203$
$e_{A3}(-\underline{r}_{A3},0,0)$; $e_{B3}(2\underline{\sigma}+\underline{r}_{A3},0,0)$	$s_{A1,A2} = 0.25$
$n_1 = n_{A3} = 2$; $n_{A1} = n_{A2} = 1$	$s_{1,A3} = (\sqrt{2}/4)\Big/\sqrt{1+\dfrac{\sigma}{(\underline{a}^2+\underline{\sigma}^2)^{1/2}}}$
$\underline{E}_b = -2.39104$	
$\underline{\sigma} = 1.04502$; $\underline{a} = 1.32709$	$Z_1 = Z_2 = Z_A - 2s_{1,A1} - s_{1,A3}$
$\underline{S}_{EiA} = 29.3377$ (Lide, 2003)	$Z_{1M} = Z_{2M} = Z_A - 2 - s_{1,A3}$
$\underline{D}_0^0 = 0.1643$ (Huber and Hertzberg, 1979)	$Z_{A1} = Z_{A2} = Z_A - s_{A1,A2} - 2s_{A1,1}$
	$Z_{A3} = Z_{B3} = Z_1 = Z_2$; $\underline{r}_{A3} = 3/Z_{A3}$
	$Z_{nA} = Z_{nB} = Z_A - 2 - \dfrac{(2\underline{\sigma})^2}{(2\underline{\sigma}+\underline{r}_{A3})^2}$

beryllium atom (see Table 1.4). The coefficient $s_{1,A3}$ is calculated with Eq. (1.59). This calculation reduces to the evaluation of the angle between the two lines which pass, respectively, through nucleus n_A and electron e_1 and through nucleus n_A and electron e_{A3}, with the aid of the coordinates of the particles, which are given in Table 2.4. The average normalized distance between the e_{A3} electron and the nucleus is calculated with the aid of the relation (B.15) from Volume I.

The normalized value of the total energy is calculated with the aid of Eq. (2.77), taking into account the symmetry of the system (namely, $\underline{E}_1 = \underline{E}_2$, $\underline{E}_{A3} = \underline{E}_{B3}$, $\underline{E}_{A1} = \underline{E}_{A2} = \underline{E}_{B1} = \underline{E}_{B2}$, $\underline{U}_{e_1 n_B} = \underline{U}_{e_2 n_A}$, $\underline{U}_{e_{A3} e_2} = \underline{U}_{e_{B3} e_1}$, and $\underline{U}_{e_{A3} n_B} = \underline{U}_{e_{B3} n_A}$):

$$\underline{E} = 2\underline{E}_1 + 2\underline{E}_{A3} + 4\underline{E}_{A1} + \underline{U}_{e_1 e_2} + 2\underline{U}_{e_1 n_B} + 2\underline{U}_{e_{A3} e_2} + 2\underline{U}_{e_{A3} n_B} + \underline{U}_{e_{A3} e_{B3}}$$

$$+ \underline{U}_{n_A n_B} + 2\underline{E}_{m1s} = -2\frac{Z_1^2}{n_1^2} - 2\frac{Z_{A3}^2}{n_{A3}^2} - 4\frac{Z_{A1}^2}{n_{A1}^2}$$

$$+ \frac{1}{2\left[\left(\frac{1}{4}\underline{\sigma}\right)^2 + \left(\frac{3}{4}\underline{a}\right)^2\right]^{1/2}} - \frac{2(Z_A-2)}{\left[\left(\frac{5}{4}\underline{\sigma}\right)^2 + \left(\frac{3}{4}\underline{a}\right)^2\right]^{1/2}} + \frac{2}{\left[\left(\frac{5}{4}\underline{\sigma}+\underline{r}_{A3}\right)^2 + \left(\frac{3}{4}\underline{a}\right)^2\right]^{1/2}}$$

$$- \frac{2(Z_A-2)}{2\underline{\sigma}+\underline{r}_{A3}} + \frac{1}{2\underline{\sigma}+2\underline{r}_{A3}} + \frac{(Z_A-2)^2}{2\underline{\sigma}} + 2\frac{Z_{A1}^{3/2}}{8n_{A1}^3}$$

$$(2.83)$$

In this case, $\underline{E}_{exp}$ is obtained from Eq. (2.80), where S_{EiA} is taken from Lide (2003), but the value of D_0^0 is obtained from figure 6.2 of Slater (1963), because Huber and Hertzberg (1979) do not have this data. The values of $\underline{E}$, $\underline{E}_{HF}$, and $\underline{E}_{exp}$ are given in Table 2.1, while r_0 and r_e are given in Table 2.2. The calculations are made with the aid of the Mathematica 7 program which is given in Section B.2.2.

2.5.3 The B$_2$ Molecule

The structure of this molecule is shown in Figure 2.7B. This molecule contains four electrons, denoted by e_{A3}, e_{A4}, e_{B3}, and e_{B4}, which do not participate to the bond. All the data necessary in calculations are given in Table 2.5. Using notations from Figure 2.7B and Eq. (B.14) from Volume I, we have $\underline{r} = \underline{r}_{A3} = \underline{r}_{A4} = (2/Z_{A3})(1 + e^2/2)$, where e is the eccentricity of the valence C_a curves of boron.

The geometrical structure of the molecule results from the system (2.70)–(2.72), while the normalized value of the total energy is calculated with the aid of Eq. (2.77). Taking into account the symmetry of

Table 2.5 Data for B$_2$ Molecule	
Particle Coordinates; Quantum Numbers. Calculated and Experimental Values	**Screening Coefficients; Order Numbers**
$n_A(0,0,0)$; $n_B(2\underline{\sigma},0,0)$	$s_{1,A1} = s_{A3,A1} = 0.862983$
$e_1\left(\frac{3}{4}\underline{\sigma}, \frac{3}{4}\underline{a}, 0\right)$; $e_2\left(\frac{5}{4}\underline{\sigma}, -\frac{3}{4}\underline{a}, 0\right)$	$s_{A1,1} = s_{A1,A3} = 0.00246735$
$e_{A3}(-\underline{d}, 0, \underline{b})$; $e_{A4}(-\underline{d}, 0, -\underline{b})$	$s_{A1,A2} = 0.25$
$e_{B3}(2\underline{\sigma} + \underline{d}, 0, -\underline{b})$; $e_{B4}(2\underline{\sigma} + \underline{d}, 0, \underline{b})$	$s_{1,A3} = (\sqrt{2}/4)/\sqrt{1 + \dfrac{\underline{\sigma}\cos\alpha}{(\underline{a}^2 + \underline{\sigma}^2)^{1/2}}}$
$n_1 = n_{A3} = n_{A4} = 2$; $n_{A1} = n_{A2} = 1$	$s_{A3,A4} = (1/4)/\sin\alpha$
$\underline{E}_b = -3.857$	$Z_1 = Z_2 = Z_A - 2s_{1,A1} - 2s_{1,A3}$
$\underline{\sigma} = 0.844851$; $\underline{a} = 1.07468$	$Z_{1M} = Z_{2M} = Z_A - 2 - 2s_{1,A3}$
$\alpha = 63°$	$Z_{A1} = Z_{A2} = Z_A - s_{A1,A2} - 3s_{A1,1}$
$\underline{S}_{EiA} = 49.3177$ (Lide, 2003)	$Z_{A3} = Z_A - 2s_{1,A1} - s_{1,A3} - s_{A3,A4}$; $Z_{A3} = Z_{A4} = Z_{B3} = Z_{B4}$
$\underline{D}_0^0 = 0.221972$ (Huber and Hertzberg, 1979)	$\underline{r} = \frac{2}{Z_{A3}}\left(1 + \frac{e^2}{2}\right)$ where $e = 0.97$ for B atom
	$\underline{d} = \underline{r}\cos\alpha$; $\underline{b} = \underline{r}\sin\alpha$
	$Z_{nA} = Z_{nB} = Z_A - 2 - \dfrac{2(2\underline{\sigma})^2(2\underline{\sigma} + \underline{d})}{[(2\underline{\sigma} + \underline{d})^2 + \underline{b}^2]^{3/2}}$

the system (namely, $\underline{E}_1 = \underline{E}_2$, $\underline{E}_{A3} = \underline{E}_{A4} = \underline{E}_{B3} = \underline{E}_{B4}$, $\underline{E}_{A1} = \underline{E}_{A2} = \underline{E}_{B1} = \underline{E}_{B2}$, $\underline{U}_{e_1 n_B} = \underline{U}_{e_2 n_A}$, $\underline{U}_{e_{A3}e_2} = \underline{U}_{e_{A4}e_2} = \underline{U}_{e_{B3}e_1} = \underline{U}_{e_{B4}e_1}$, $\underline{U}_{e_{A3}n_B} = \underline{U}_{e_{A4}n_B} = \underline{U}_{e_{B3}n_A} = \underline{U}_{e_{B4}n_A}$, $\underline{U}_{e_{A3}e_{B3}} = \underline{U}_{e_{A4}e_{B4}}$, and $\underline{U}_{e_{A3}e_{B4}} = \underline{U}_{e_{A4}e_{B3}}$), the expression of the energy becomes:

$$\underline{E} = 2\underline{E}_1 + 4\underline{E}_{A3} + 4\underline{E}_{A1} + \underline{U}_{e_1 e_2} + 2\underline{U}_{e_1 n_B} + 4\underline{U}_{e_{A3}e_2} + 4\underline{U}_{e_{A3}n_B}$$

$$+ 2\underline{U}_{e_{A3}e_{B3}} + 2\underline{U}_{e_{A3}e_{B4}} + \underline{U}_{n_A n_B} + 2\underline{E}_{m1s} = -2\frac{Z_1^2}{n_1^2} - 4\frac{Z_{A3}^2}{n_{A3}^2} - 4\frac{Z_{A1}^2}{n_{A1}^2}$$

$$+ \frac{1}{2\left[\left(\frac{1}{4}\underline{\sigma}\right)^2 + \left(\frac{3}{4}a\right)^2\right]^{1/2}} - \frac{2(Z_A - 2)}{\left[\left(\frac{5}{4}\underline{\sigma}\right)^2 + \left(\frac{3}{4}a\right)^2\right]^{1/2}} + \frac{4}{\left[\left(\underline{d} + \frac{5}{4}\underline{\sigma}\right)^2 + \left(\frac{3}{4}a\right)^2 + \underline{b}^2\right]^{1/2}}$$

$$- \frac{4(Z_A - 2)}{[(\underline{d} + 2\underline{\sigma})^2 + \underline{b}^2]^{1/2}} + \frac{1}{[(\underline{d} + \underline{\sigma})^2 + \underline{b}^2]^{1/2}} + \frac{1}{\underline{\sigma} + \underline{d}} + \frac{(Z_A - 2)^2}{2\underline{\sigma}} + 2\frac{Z_{A1}^{3/2}}{8n_{A1}^3}$$

(2.84)

The angle α is a parameter. We perform the calculation for a lot of values of this angle and choose the value for which the total energy is minimum. This value is given in Table 2.5. This value is very easy to find because the value of 2α is close to the angle between the axes of the C_a curves for the valence electrons in the boron atom, that is equal to $120°$.

In this case and in all the cases which follow, $\underline{E}_{\exp}$ is obtained with the aid of Eq. (2.80), where S_{EiA} and D_0^0 are taken, respectively, from Lide (2003) and Huber and Hertzberg (1979). All the calculated data are given in Table 2.5. The values of $\underline{E}$, $\underline{E}_{HF}$, and $\underline{E}_{\exp}$ are shown in Table 2.1 while r_0 and r_e are given in Table 2.2. The calculations are made with a Mathematica 7 program, which is given in Section B.2.3.

2.5.4 The C$_2$ Molecule

We analyze the C_2 molecule in three cases: when the bond is single, double, and triple. We calculate the total energy and choose the case having the lowest total energy. We consider that this case corresponds to the real C_2 molecule.

2.5.4.1 The C$_2$ Molecule with Single Bond

The structure of this molecule is shown in Figure 2.8A. This molecule contains six electrons, denoted by e_{A3}, e_{A4}, e_{A5}, e_{B3}, e_{B4}, and e_{B5}, which

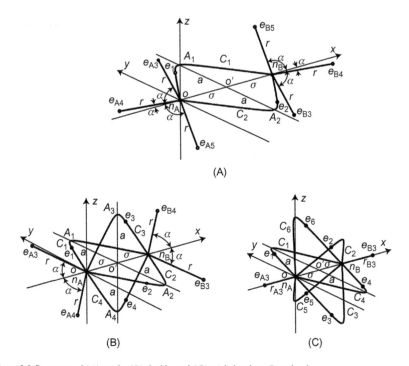

Figure 2.8 Structures of (A) single, (B) double, and (C) triple bonds in C_2 molecule.

do not participate to the bond. The coordinates of these electrons are as follows:

$$e_1\left(\frac{3}{4}\underline{\sigma}, \frac{3}{4}\underline{a}, 0\right), \quad e_2\left(\frac{5}{4}\underline{\sigma}, -\frac{3}{4}\underline{a}, 0\right), \quad e_{A3}(-\underline{d}, 0, \underline{b}),$$

$$e_{A4}\left(-\underline{d}, \frac{\sqrt{3}}{2}\underline{b}, -\frac{1}{2}\underline{b}\right), \quad e_{A5}\left(-\underline{d}, -\frac{\sqrt{3}}{2}\underline{b}, -\frac{1}{2}\underline{b}\right),$$

$$e_{B3}(2\underline{\sigma} + \underline{d}, 0, -\underline{b}), \quad e_{B4}\left(2\underline{\sigma} + \underline{d}, -\frac{\sqrt{3}}{2}\underline{b}, \frac{1}{2}\underline{b}\right), \quad \text{and}$$

$$e_{B5}\left(2\underline{\sigma} + \underline{d}, \frac{\sqrt{3}}{2}\underline{b}, \frac{1}{2}\underline{b}\right)$$

A calculation similar to that for the B_2 molecule leads to the following values: $\underline{E}_b = -5.69164$, $\underline{\sigma} = 0.751248$, $\underline{a} = 0.857921$, $r_0 = 1.59018A$, and $\underline{E} = -150.365$. The calculation is made with a Mathematica 7 program, which is given in Section B.2.4.

2.5.4.2 The C_2 Molecule with Double Bond

The structure of this molecule is shown in Figure 2.8B. This molecule contains four electrons, denoted by e_{A3}, e_{A4}, e_{B3}, and e_{B4}, which do not participate to the bond. The geometrical structure of the molecule results from the system (2.31)–(2.33), which can be written in normalized form, as follows:

$$\underline{E}_b = -4\frac{Z_1^2}{n_1^2} - \frac{16(Z_{1M} + s_{1,2})}{\sqrt{9\underline{a}^2 + 25\underline{\sigma}^2}} + \frac{8}{\sqrt{4.5\underline{a}^2 + \underline{\sigma}^2}} + \frac{Z_{nA}^2}{2\underline{\sigma}} \tag{2.85}$$

$$\underline{E}_b = -\frac{4n_1\sqrt{Z_{1M}}}{(\underline{a}^2 + \underline{\sigma}^2)^{3/4}} \tag{2.86}$$

$$\underline{E}_b = -\frac{8(Z_{1M} + s_{1,2})}{\sqrt{\underline{a}^2 + \underline{\sigma}^2}} + \frac{1}{\underline{a}} + \frac{2\sqrt{2}}{\underline{a}} + \frac{Z_{nA}^2}{2\underline{\sigma}} \tag{2.87}$$

All the data necessary for calculations, together with the results of the calculations are given in Table 2.6. The angle α corresponds to the

Table 2.6 Data for C_2 Molecule with Double Bond	
Particle Coordinates; Quantum Numbers. Calculated and Experimental Values	**Screening Coefficients; Order Numbers**
$n_A(0,0,0)$; $n_B(2\underline{\sigma},0,0)$	$s_{1,A1} = s_{A3,A1} = s_{A4,A1} = 0.85505$
$e_1\left(\frac{3}{4}\underline{\sigma}, \frac{3}{4}\underline{a}, 0\right)$; $e_2\left(\frac{3}{4}\underline{\sigma}, -\frac{3}{4}\underline{a}, 0\right)$; $e_3\left(\frac{5}{4}\underline{\sigma}, 0, \frac{3}{4}\underline{a}\right)$; $e_4\left(\frac{5}{4}\underline{\sigma}, 0, -\frac{3}{4}\underline{a}\right)$	$s_{A1,1} = s_{A1,A3} = s_{A1,A4} = 0.00269287$
$e_{A3}\left(-\underline{d}, \frac{\sqrt{2}}{2}\underline{b}, \frac{\sqrt{2}}{2}\underline{b}\right)$; $e_{A4}\left(-\underline{d}, -\frac{\sqrt{2}}{2}\underline{b}, -\frac{\sqrt{2}}{2}\underline{b}\right)$	$s_{A1,A2} = 0.25$; $s_{1,2} = \sqrt{\underline{a}^2 + \underline{\sigma}^2}/(4\underline{a})$
$e_{B3}\left(2\underline{\sigma} + \underline{d}, -\frac{\sqrt{2}}{2}\underline{b}, -\frac{\sqrt{2}}{2}\underline{b}\right)$; $e_{B4}\left(2\underline{\sigma} + \underline{d}, \frac{\sqrt{2}}{2}\underline{b}, \frac{\sqrt{2}}{2}\underline{b}\right)$	$s_{1,A3} = s_{2,A4} = \dfrac{\sqrt{2}}{4\left(1 + \dfrac{\underline{\sigma}\cos\alpha - \frac{\sqrt{2}}{2}\underline{a}\sin\alpha}{(\underline{a}^2 + \underline{\sigma}^2)^{1/2}}\right)^{1/2}}$
$n_1 = n_{A3} = n_{A4} = 2$; $n_{A1} = n_{A2} = 1$	$s_{1,A4} = s_{2,A3} = \dfrac{\sqrt{2}}{4\left(1 + \dfrac{\underline{\sigma}\cos\alpha + \frac{\sqrt{2}}{2}\underline{a}\sin\alpha}{(\underline{a}^2 + \underline{\sigma}^2)^{1/2}}\right)^{1/2}}$
$\underline{E}_b = -12.1816$	$s_{A3,A4} = 1/(4\sin\alpha)$
$\underline{\sigma} = 0.641059$; $\underline{a} = 0.894043$	$Z_1 = Z_2 = Z_A - 2s_{1,A1} - s_{1,2} - s_{1,A3} - s_{1,A4}$
$\alpha = 41°$	$Z_{1M} = Z_{2M} = Z_A - 2 - s_{1,2} - s_{1,A3} - s_{1,A4}$
$\underline{S}_{EiA} = 75.7133$ (Lide, 2003)	$Z_{A1} = Z_{A2} = Z_A - s_{A1,A2} - 4s_{A1,1}$
$\underline{D}_0^0 = 0.45644$ (Huber and Hertzberg, 1979)	$Z_{A3} = Z_A - 2s_{1,A1} - s_{1,A3} - s_{1,A4} - s_{A3,A4}$
	$Z_{A3} = Z_{A4} = Z_{B3} = Z_{B4}$
	$\underline{r} = \frac{2}{Z_{A3}}\left(1 + \frac{e^2}{2}\right)$
	where $e = 0.98$ for C atom
	$\underline{d} = \underline{r}\cos\alpha$; $\underline{b} = \underline{r}\sin\alpha$
	$Z_{nA} = Z_{nB} = Z_A - 2 - \dfrac{2(2\underline{\sigma})^2(2\underline{\sigma} + \underline{d})}{[(2\underline{\sigma} + \underline{d})^2 + \underline{b}^2]^{3/2}}$

minimum value of the total energy. Using the notations from Figure 2.8B and Eq. (B.14) from Volume I, we have $\underline{r} = \underline{r}_{A3} = \underline{r}_{A4} = (3/Z_{A3})(1 + e^2/2)$, where e is the eccentricity of the valence C_a curves of carbon. The normalized value of the total energy is calculated with the aid of Eq. (2.77). Taking into account the symmetry of the system (namely,

$\underline{E}_1 = \underline{E}_2 = \underline{E}_3 = \underline{E}_4$, $\underline{E}_{A3} = \underline{E}_{A4} = \underline{E}_{B3} = \underline{E}_{B4}$, $\underline{E}_{A1} = \underline{E}_{A2} = \underline{E}_{B1} = \underline{E}_{B2}$, $\underline{U}_{e_1e_3} = \underline{U}_{e_1e_4} = \underline{U}_{e_2e_3} = \underline{U}_{e_2e_4}$, $\underline{U}_{e_1n_B} = \underline{U}_{e_2n_B} = \underline{U}_{e_3n_A} = \underline{U}_{e_4n_A}$, $\underline{U}_{e_{A3}e_3} = \underline{U}_{e_{A4}e_4} = \underline{U}_{e_{B3}e_2} = \underline{U}_{e_{B4}e_1}$, $\underline{U}_{e_{A3}e_4} = \underline{U}_{e_{A4}e_3} = \underline{U}_{e_{B3}e_1} = \underline{U}_{e_{B4}e_2}$, $\underline{U}_{e_{A3}n_B} = \underline{U}_{e_{A4}n_B} = \underline{U}_{e_{B3}n_A} = \underline{U}_{e_{B4}n_A}$, $\underline{U}_{e_{A3}e_{B3}} = \underline{U}_{e_{A4}e_{B4}}$, and $\underline{U}_{e_{A3}e_{B4}} = \underline{U}_{e_{A4}e_{B3}}$), this relation can be written for C_2 molecule with double bond, as follows:

$$\underline{E} = 4\underline{E}_1 + 4\underline{E}_{A3} + 4\underline{E}_{A1} + 4\underline{U}_{e_1e_3} + 4\underline{U}_{e_1n_B} + 4\underline{U}_{e_{A3}n_B} + 2\underline{U}_{e_{A3}e_{B3}}$$

$$+ 2\underline{U}_{e_{A3}e_{B4}} + 4\underline{U}_{e_{A3}e_3} + 4\underline{U}_{e_{A3}e_4} + \underline{U}_{n_An_B} + 2\underline{E}_{mls} = -4\frac{Z_1^2}{n_1^2} - 4\frac{Z_{A3}^2}{n_{A3}^2}$$

$$-4\frac{Z_{A1}^2}{n_{A1}^2} + \frac{4}{\left[\left(\frac{1}{2}\underline{\sigma}\right)^2 + 2\left(\frac{3}{4}\underline{a}\right)^2\right]^{1/2}} - \frac{4(Z_A - 2)}{\left[\left(\frac{5}{4}\underline{\sigma}\right)^2 + \left(\frac{3}{4}\underline{a}\right)^2\right]^{1/2}} - \frac{4(Z_A - 2)}{[(\underline{d} + 2\underline{\sigma})^2 + \underline{b}^2]^{1/2}}$$

$$+ \frac{1}{[(\underline{d} + \underline{\sigma})^2 + \underline{b}^2]^{1/2}} + \frac{1}{\underline{\sigma} + \underline{d}} + \frac{4}{\left[\left(\underline{d} + \frac{5}{4}\underline{\sigma}\right)^2 + \left(\frac{\sqrt{2}}{2}\underline{b}\right)^2 + \left(\frac{3}{4}\underline{a} - \frac{\sqrt{2}}{2}\underline{b}\right)^2\right]^{1/2}}$$

$$+ \frac{4}{\left[\left(\underline{d} + \frac{5}{4}\underline{\sigma}\right)^2 + \left(\frac{\sqrt{2}}{2}\underline{b}\right)^2 + \left(\frac{3}{4}\underline{a} + \frac{\sqrt{2}}{2}\underline{b}\right)^2\right]^{1/2}} + \frac{(Z_A - 2)^2}{2\underline{\sigma}} + 2\frac{Z_{A1}^{3/2}}{8n_{A1}^3}$$

(2.88)

We have obtained the values $\underline{E} = -151.148$ and $r_0 = 1.35694A$. The calculations were made with the aid of the Mathematica 7 program which is given in Section B.2.5.

2.5.4.3 The C_2 Molecule with Triple Bond

The structure of this molecule is shown in Figure 2.8C. This molecule contains two electrons, denoted by e_{A3} and e_{B3}, which do not participate to the bond. The geometrical structure of the molecule results from the system (2.34)–(2.36), which can be written in normalized form, as follows:

$$\underline{E}_b = -6\frac{Z_1^2}{n_1^2} - \frac{24(Z_{1M} + 2s_{1,2})}{\sqrt{9\underline{a}^2 + 25\underline{\sigma}^2}} + \frac{6}{\sqrt{9\underline{a}^2 + \underline{\sigma}^2}} + \frac{24}{\sqrt{9\underline{a}^2 + 4\underline{\sigma}^2}} + \frac{Z_{nA}^2}{2\underline{\sigma}}$$

$$(2.89)$$

$$\underline{E}_b = -\frac{6n_1\sqrt{Z_{1M}}}{(\underline{a}^2 + \underline{\sigma}^2)^{3/4}} \tag{2.90}$$

$$\underline{E}_b = -\frac{12(Z_{1M} + 2s_{1,2})}{\sqrt{\underline{a}^2 + \underline{\sigma}^2}} + \frac{6}{\sqrt{3\underline{a}}} + \frac{6}{\underline{a}} + \frac{3}{2\underline{a}} + \frac{Z_{nA}^2}{2\underline{\sigma}} \tag{2.91}$$

All the theoretical data are given in Table 2.7. Taking into account the symmetry of the system (namely, $\underline{E}_1 = \underline{E}_2 = \cdots = \underline{E}_6$, $\underline{E}_{A3} = \underline{E}_{B3}$, $\underline{E}_{A1} = \underline{E}_{A2} = \underline{E}_{B1} = \underline{E}_{B2}$, $\underline{U}_{e_1e_4} = \underline{U}_{e_2e_5} = \underline{U}_{e_3e_6}$, $\underline{U}_{e_1e_5} = \underline{U}_{e_5e_3} = \underline{U}_{e_3e_4} = \underline{U}_{e_4e_2} = \underline{U}_{e_2e_6} = \underline{U}_{e_6e_1}$, $\underline{U}_{e_{A3}n_B} = \underline{U}_{e_{B3}n_A}$, $\underline{U}_{e_in_B} = \underline{U}_{e_jn_A}$, and $\underline{U}_{e_{A3}e_j} =$

Table 2.7 Data for C_2 Molecule with Triple Bond	
Particle Coordinates; Quantum Numbers. Calculated and Experimental Values	Screening Coefficients; Order Numbers
$n_A(0,0,0)$; $n_B(2\underline{\sigma},0,0)$	$s_{1,A1} = s_{A3,A1} = 0.85505$
$e_1\left(\frac{3}{4}\underline{\sigma},\frac{3}{4}\underline{a},0\right)$; $e_2\left(\frac{3}{4}\underline{\sigma}, -\frac{3}{4}\cdot\frac{1}{2}\underline{a},\frac{3}{4}\cdot\frac{\sqrt{3}}{2}\underline{a}\right)$; $e_3\left(\frac{3}{4}\underline{\sigma}, -\frac{3}{4}\cdot\frac{1}{2}\underline{a}, -\frac{3}{4}\cdot\frac{\sqrt{3}}{2}\underline{a}\right)$; $e_4\left(\frac{5}{4}\underline{\sigma}, -\frac{3}{4}\underline{a},0\right)$; $e_5\left(\frac{5}{4}\underline{\sigma},\frac{3}{4}\cdot\frac{1}{2}\underline{a}, -\frac{3}{4}\cdot\frac{\sqrt{3}}{2}\underline{a}\right)$; $e_6\left(\frac{5}{4}\underline{\sigma},\frac{3}{4}\cdot\frac{1}{2}\underline{a},\frac{3}{4}\cdot\frac{\sqrt{3}}{2}\underline{a}\right)$	$s_{A1,1} = s_{A1,A3} = 0.00269287$
$e_{A3}(-\underline{r}_{A3},0,0)$; $e_{A4}(2\underline{\sigma} + \underline{r}_{A3},0,0)$	$s_{A1,A2} = 0.25$
$n_1 = n_{A3} = 2$; $n_{A1} = n_{A2} = 1$	$s_{1,2} = s_{2,3} = s_{1,3} = \dfrac{\sqrt{2}}{\sqrt[4]{1 - \dfrac{\underline{\sigma}^2 - (1/2)\underline{a}^2}{\underline{\sigma}^2 + \underline{a}^2}}}$
$\underline{E}_b = -17.1499$	$s_{1,A3} = s_{2,A3} = s_{3,A3} = \dfrac{\sqrt{2}}{\sqrt[4]{1 + \dfrac{\underline{\sigma}}{\sqrt{\underline{\sigma}^2 + \underline{a}^2}}}}$
$\underline{\sigma} = 0.639379$; $\underline{a} = 0.942817$	$Z_1 = Z_2 = Z_3 = Z_A - 2s_{1,A1} - 2s_{1,2} - s_{1,A3}$
	$Z_{1M} = Z_{2M} = Z_{3M} = Z_A - 2 - 2s_{1,2} - s_{1,A3}$
	$Z_{A1} = Z_{A2} = Z_A - s_{A1,A2} - 4s_{A1,1}$
	$Z_{A3} = Z_{B3} = Z_A - 2s_{1,A1} - 3s_{1,A3}$
	$\underline{r}_{A3} = \frac{2}{Z_{A3}}\left(1 + \frac{e^2}{2}\right)$
	where $e = 0.98$ for C atom
	$Z_{nA} = Z_{nB} = Z_A - 2 - \dfrac{(2\underline{\sigma})^2}{(2\underline{\sigma} + \underline{r}_{A3})^2}$

$\underline{U}_{e_{B3}e_i}$, where $i = 1, 2, 3$ and $j = 4, 5, 6$), the total energy results from the following relation:

$$\underline{E} = 6\underline{E}_1 + 2\underline{E}_{A3} + 4\underline{E}_{A1} + 3\underline{U}_{e_1e_4} + 6\underline{U}_{e_1e_5} + 6\underline{U}_{e_1n_B} + 6\underline{U}_{e_{A3}e_4}$$

$$+ 2\underline{U}_{e_{A3}n_B} + \underline{U}_{e_{A3}e_{B_3}} + \underline{U}_{n_An_B} + 2\underline{E}_{mls} = -6\frac{Z_1^2}{n_1^2} - 2\frac{Z_{A3}^2}{n_{A3}^2} - 4\frac{Z_{A1}^2}{n_{A1}^2}$$

$$+ \frac{3}{\left[\left(\frac{1}{2}\underline{\sigma}\right)^2 + \left(\frac{3}{2}\underline{a}\right)^2\right]^{1/2}} + \frac{6}{\left[\left(\frac{1}{2}\underline{\sigma}\right)^2 + \left(\frac{3}{4}\underline{a}\right)^2\right]^{1/2}} - \frac{6(Z_A - 2)}{\left[\left(\frac{5}{4}\underline{\sigma}\right)^2 + \left(\frac{3}{4}\underline{a}\right)^2\right]^{1/2}}$$

$$- \frac{2(Z_A - 2)}{\underline{r}_{A3} + 2\underline{\sigma}} + \frac{6}{\left[\left(\underline{r}_{A3} + \frac{5}{4}\underline{\sigma}\right)^2 + \left(\frac{3}{4}\underline{a}\right)^2\right]^{1/2}} + \frac{1}{2\underline{r}_{A3} + 2\underline{\sigma}} + \frac{(Z_A - 2)^2}{2\underline{\sigma}} + 2\frac{Z_{A1}^{3/2}}{8n_{A1}^3}$$

$$(2.92)$$

We have obtained the values $\underline{E} = -150.923$ and $r_0 = 1.35339A$. The calculations were made with the aid of the Mathematica 7 program which is given in Section B.2.6.

If we compare the three above values of the total energy, we will see that the energy corresponding to the double bond is minimum, resulting that the real bond of the C_2 molecule is double. This result is identical to that obtained by pure quantum evaluations (see, e.g., page 97 of Coulson (1961)). In Table 2.1, it is shown the value of $\underline{E}$ corresponding to the C_2 molecule with double bond, together with $\underline{E}_{HF}$ and $\underline{E}_{exp}$. The value of r_0 from Table 2.2 corresponds also to C_2 molecule with double bond.

A comparison between, on one hand, the symmetry properties of the systems composed of the C atoms and the electrons which do not participate to the bond (Figure 2.8A−C), and, on the other hand, the experimental symmetry properties of the ethane, ethylene, and acetylene molecules, namely, C_2H_6, C_2H_4, and C_2H_2 (see CCCBDB), show that these properties are identical. For example, all these structures have a center of symmetry, the configurations of C_2H_4 and of n_A, n_B, e_{A3}, e_{A4}, e_{B3}, and e_{B4}, in the case of double bond, are plane, or the configurations of C_2H_2 and of n_A, n_B, e_{A3}, and e_{B3}, in the case of triple bond, are linear. This explains why adding 6, 4 and, respectively, 2 hydrogen atoms to the structures of C_2 with single, double, and triple bonds leads to the formation of the C_2H_6, C_2H_4 and, respectively, C_2H_2 molecules.

Our calculations explain also the property that the length of the bond decreases with the increasing of the bond order.

2.5.5 The LiH Molecule

We analyze now the following molecules: LiH, BeH, BH, and CH. The model of ionic bond is in agreement with the experimental data for LiH, BeH, while the model of covalent bond is in agreement with the experimental data for BH and CH. For example, if we apply the covalent model in the case of LiH, it will lead to a symmetrical molecule, for which $\sigma_1 = \sigma_2$, which is in strong disagreement with the fact that this molecule has a big electrical dipole moment.

The structure of the LiH molecule is shown in Figure 2.4. This molecule contains only valence electrons that participate to the bond. The geometrical structure of the molecule results from the system (2.45), (2.46), (2.53), (2.54), and (2.57), which can be written in normalized form, as follows:

$$\underline{E}_b = -2\frac{Z_{1(A)}^2}{n_{1(A)}^2} - \frac{8[Z_{1M(B)} + s_{1,2(B)}]}{\sqrt{9\underline{a}^2 + (\underline{\sigma}_1 + 4\underline{\sigma}_2)^2}} + \frac{Z_{nA}Z_{nB}}{\underline{\sigma}_1 + \underline{\sigma}_2} \tag{2.93}$$

$$\underline{E}_b = -2\frac{Z_{1(B)}^2}{n_{1(B)}^2} - \frac{8[Z_{1M(A)} + s_{1,2(A)}]}{\sqrt{9\underline{a}^2 + (\underline{\sigma}_2 + 4\underline{\sigma}_1)^2}} + \frac{Z_{nA}Z_{nB}}{\underline{\sigma}_1 + \underline{\sigma}_2} \tag{2.94}$$

$$\underline{E}_b = -\frac{2n_{1(A)}\sqrt{Z_{1M(A)}}}{(\underline{\sigma}_1^2 + \underline{a}^2)^{3/4}t_{cA}} \tag{2.95}$$

$$\underline{E}_b = -\frac{2n_{1(B)}\sqrt{Z_{1M(B)}}}{(\underline{\sigma}_2^2 + \underline{a}^2)^{3/4}t_{cB}} \tag{2.96}$$

$$\underline{E}_b = -\frac{2[Z_{1M(A)} + s_{1,2(A)}]}{\sqrt{\underline{\sigma}_1^2 + \underline{a}^2}} - \frac{2[Z_{1M(B)} + s_{1,2(B)}]}{\sqrt{\underline{\sigma}_2^2 + \underline{a}^2}} + \frac{1}{2\underline{a}} + \frac{Z_{nA}Z_{nB}}{\underline{\sigma}_1 + \underline{\sigma}_2} + 2\underline{T}_m \tag{2.97}$$

where

$$t_{cA} = \frac{2}{\pi a_{cA}^{3/2}}(\arcsin\sqrt{a_{cA}} - \sqrt{a_{cA}}\sqrt{1 - a_{cA}}) \tag{2.98}$$

$$a_{cA} = 1 - \frac{\underline{T}_m\sqrt{\underline{\sigma}_1^2 + \underline{a}^2}}{Z_{1M(A)}} \tag{2.99}$$

$$t_{cB} = \frac{2}{\pi a_{cB}^{3/2}} (\arcsin \sqrt{a_{cB}} - \sqrt{a_{cB}} \sqrt{1 - a_{cB}}) \qquad (2.100)$$

$$a_{cB} = 1 - \frac{T_m \sqrt{\sigma_2^2 + a^2}}{Z_{1M(B)}} \qquad (2.101)$$

The system (2.93)–(2.97) has five equations with five unknowns, which are E_b, σ_1, σ_2, a, and T_m.

We follow again the algorithm presented in Section 2.4 and calculate the screening coefficients, order numbers and solve the system (2.93)–(2.97). All these values are given in Table 2.8. The screening coefficients between the 1s electrons and the e_1 and e_2 electrons, namely $s_{1,A1}$ and $s_{A1,1}$, when the system is in the phase A are identical, respectively to s_{31e} and s_{13e}, which have been calculated in Chapter 1 for the $1s^2 2s$ state of lithium atom (see Table 1.4). The coefficients $s_{A1,A2}$, $s_{1,2(A)}$, and $s_{1,2(B)}$ are calculated with Eq. (1.59). The principal quantum numbers corresponding to e_1 and e_2 electrons, in phases A and B, are $n_{1(A)} = n_{2(A)} = 2$ and $n_{1(B)} = n_{2(B)} = 1$.

Table 2.8 Data for LiH Molecule	
Particle Coordinates; Quantum Numbers. Calculated and Experimental Values	**Screening Coefficients; Order Numbers**
$n_A(0,0,0)$; $n_B(\sigma_1 + \sigma_2, 0, 0)$	$s_{1,A1} = 0.854942$
Phase A:	$s_{A1,1} = 0.0013792$
$e_1 \left(\frac{3}{4}\sigma_1, \frac{3}{4}a, 0\right)$; $e_2 \left(\frac{3}{4}\sigma_1, -\frac{3}{4}a, 0\right)$	$s_{A1,A2} = 0.25$
Phase B:	$s_{1,2(A)} = \sqrt{\sigma_1^2 + a^2}/(4a)$
$e_1 \left(\sigma_1 + \frac{1}{4}\sigma_2, \frac{3}{4}a, 0\right)$; $e_2 \left(\sigma_1 + \frac{1}{4}\sigma_2, -\frac{3}{4}a, 0\right)$	$s_{1,2(B)} = \sqrt{\sigma_2^2 + a^2}/(4a)$
$n_{1(A)} = n_{2(A)} = 2$; $n_{1(B)} = n_{2(B)} = 1$; $n_{A1} = n_{A2} = 1$	$Z_{1(A)} = Z_A - 2s_{1,A1} - s_{1,2(A)}$
$E_b = -1.65659$;	$Z_{1M(A)} = Z_A - 2 - s_{1,2(A)}$
$\sigma_1 = 1.47713$; $\sigma_2 = 0.267425$	$Z_{A1} = Z_A - s_{A1,A2} - 2s_{A1,1}$
$a = 1.18038$; $T_m = 0.0284159$	$Z_{1(B)} = Z_B - s_{1,2(B)}$
$S_{EiA} = 14.9563$ (Lide, 2003)	$Z_{1M(B)} = Z_B - s_{1,2(B)}$
$S_{EiB} = 1$	$Z_{nA} = Z_A - 2$; $Z_{nB} = Z_B$
$D_0^0 = 0.178511$ (Huber and Hertzberg, 1979)	
$\mu_e = 6.1491$ D	
$\mu_{eHF} = (4.839, 6.083)$ D (CCCBDB)	

Since in this case $\underline{E}_b$ has the significance of the total energy from which the energies of the 1s electrons are subtracted, we have:

$$E = \underline{E}_b + 2\underline{E}_{A1} + \underline{E}_{m1s} = \underline{E}_b - 2\frac{Z_{A1}^2}{n_{A1}^2} + \frac{Z_{A1}^{3/2}}{8n_{A1}^3} \tag{2.102}$$

The total energy $\underline{E}$, together with the best two values of $\underline{E}_{HF}$, obtained from CCCBDB, and the experimental value $\underline{E}_{exp}$ are presented in Table 2.1. The experimental value is given by:

$$\underline{E}_{exp} = -\underline{S}_{EiA} - \underline{S}_{EiB} - \underline{D}_0^0 \tag{2.103}$$

where $\underline{S}_{EiA}$ and $\underline{S}_{EiB}$ are the sums of the ionization energies for the two atoms. The values of $\underline{S}_{EiA}$ and $\underline{S}_{EiB}$ are taken from Lide (2003), while $\underline{D}_0^0$ is taken from Huber and Hertzberg (1979).

In this case, we have:

$$r_0 = \sigma_1 + \sigma_2 \tag{2.104}$$

The values of r_0, which is calculated with Eq. (2.104), and r_e, which is taken from CCCBDB, are given in angstroms in Table 2.2.

The electric dipole moment of the molecule is denoted by $\overline{\mu}_e$. It is calculated when the electrons are situated at the points A_1 and A_2, which are the average positions on the molecular C_a curves. We take into account the convention that the positive sense of the vector $\overline{\mu}_e$ is from the negative charge to the positive charge. In this case, the sign convention is the same as that from CCCBDB. The expression of $\overline{\mu}_e$ in the case of the LiH molecule, when the result is given in debyes (D), is:

$$\mu_e = -5.08315[-2\underline{\sigma}_1 + Z_B(\underline{\sigma}_1 + \underline{\sigma}_2)] \tag{2.105}$$

The theoretical value, calculated with Eq. (2.105), and the experimental value, taken from CCCBDB, are equal, respectively, to 6.1491 D and 5.88 D. The reference CCCBDB does not contain experimental values of the electric dipole moment in the cases of the molecules BeH, BH, and CH. In Table 2.8, we compare our theoretical values with the theoretical values of the electric dipole moment, calculated using the Hartree−Fock method, which are taken from CCCBDB. The analysis of the data from CCCBDB shows that the dispersion of the theoretical values of the electric dipole moments, in the case of a given system, is very high. For this reason, in Table 2.8, we give the domain of the theoretical values of the dipole moments, taken from CCCBDB, and computed using the Hartree−Fock method. These values are denoted by μ_{eHF}. All the calculations are made with the Mathematica 7 program, which is shown in Section B.2.7.

2.5.6 The BeH molecule

The structure of this molecule is shown in Figure 2.9A. This molecule contains one electron, denoted by e_{A3}, which does not participate to the bond. The distance r_{A3} varies slightly when the bond electrons move in the phases A and B. Since the bond electrons are in the majority of the time far from nucleus, we consider an average value, approximately equal to that in the case of the isolated atom, which is given by the relation $\underline{r}_{A3} = 3/(Z_A - 2s_{A3,A1})$. In this case, the value of E_b is slightly different in the two phases, being equal, respectively, to E_{b1} and E_{b2}, in the phases A and B. The system (2.93)−(2.97) becomes:

$$\underline{E}_{b1} = -2\frac{Z_{1(A)}^2}{n_{1(A)}^2} - \frac{8[Z_{1M(B)} + s_{1,2(B)}]}{\sqrt{9\underline{a}^2 + (\underline{\sigma}_1 + 4\underline{\sigma}_2)^2}} + \frac{Z_{nA}Z_{nB}}{\underline{\sigma}_1 + \underline{\sigma}_2} \qquad (2.106)$$

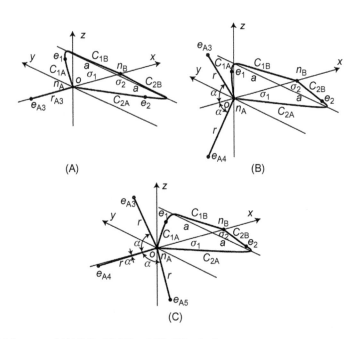

Figure 2.9 Structures of (A) BeH, (B) BH, and (C) CH molecules.

$$\underline{E}_{b2} = -2\frac{Z_{1(B)}^2}{n_{1(B)}^2} - \frac{8[Z_{1M(A)} + s_{1,2(A)}]}{\sqrt{9\underline{a}^2 + (\underline{\sigma}_2 + 4\underline{\sigma}_1)^2}} + \frac{Z_{nA}Z_{nB}}{\underline{\sigma}_1 + \underline{\sigma}_2} \qquad (2.107)$$

$$\underline{E}_{b1} = -\frac{2n_{1(A)}\sqrt{Z_{1M(A)}}}{(\underline{\sigma}_1^2 + \underline{a}^2)^{3/4} t_{cA}} \qquad (2.108)$$

$$\underline{E}_{b2} = -\frac{2n_{1(B)}\sqrt{Z_{1M(B)}}}{(\underline{\sigma}_2^2 + \underline{a}^2)^{3/4} t_{cB}} \qquad (2.109)$$

$$\frac{1}{2}(\underline{E}_{b1} + \underline{E}_{b2}) = -\frac{2[Z_{1M(A)} + s_{1,2(A)}]}{\sqrt{\underline{\sigma}_1^2 + \underline{a}^2}} - \frac{2[Z_{1M(B)} + s_{1,2(B)}]}{\sqrt{\underline{\sigma}_2^2 + \underline{a}^2}} + \frac{1}{2\underline{a}} + \frac{Z_{nA}Z_{nB}}{\underline{\sigma}_1 + \underline{\sigma}_2} + 2\underline{T}_m \qquad (2.110)$$

where t_{cA} and t_{cB} are calculated, respectively, with Eqs. (2.98) and (2.100). This system has five equations with five unknowns, which are $\underline{E}_{b1}$, $\underline{E}_{b2}$, $\underline{\sigma}_1$, $\underline{\sigma}_2$, and $\underline{a}$. The value of $\underline{T}_m$ is very small; it is chosen from the condition of minimum value of the total energy. All the parameters and expressions which are used in calculations are shown in Table 2.9. The screening coefficients between the 1s electrons and the e_1, e_2, and e_{A3} electrons are identical to s_{31e} and s_{13e}, which have been calculated in Chapter 1 for the $1s^2s^2$ state of beryllium atom (see Table 1.4). In the case of heteronuclear molecules, the values of Z_{nA} and Z_{nB} are calculated with Eq. (2.76) in which, instead of 2σ, we consider the expression $\sigma_1 + \sigma_2$.

The normalized value of the total energy is calculated with the aid of Eq. (2.77), which is written for BeH molecule, when the electrons are situated at points A_1 and A_2. Taking into account the symmetry of the system (namely, $\underline{U}_{e_1 n_A} = \underline{U}_{e_2 n_A}$ and $\underline{U}_{e_1 n_B} = \underline{U}_{e_2 n_B}$), we have:

$$\underline{E} = \underline{E}_{A3} + 2\underline{E}_{A1} + 2\underline{U}_{e_1 n_A} + 2\underline{U}_{e_1 n_B} + \underline{U}_{e_1 e_2} + \underline{U}_{e_{A3} n_B}$$
$$+ \underline{U}_{n_A n_B} + \underline{E}_{m1s} + 2\underline{T}_m = -\frac{Z_{A3}^2}{n_{A3}^2} - 2\frac{Z_{A1}^2}{n_{A1}^2} - \frac{2[Z_{1M(A)} + s_{1,2(A)}]}{\sqrt{\underline{\sigma}_1^2 + \underline{a}^2}}$$
$$- \frac{2[Z_{1M(B)} + s_{1,2(B)}]}{\sqrt{\underline{\sigma}_2^2 + \underline{a}^2}} + \frac{1}{2\underline{a}} - \frac{Z_B}{\underline{\sigma}_1 + \underline{\sigma}_2 + r_{A3}} + \frac{(Z_A - 2)Z_B}{\underline{\sigma}_1 + \underline{\sigma}_2} + \frac{Z_{A1}^{3/2}}{8n_{A1}^3} + 2\underline{T}_m \qquad (2.111)$$

Table 2.9 Data for BeH Molecule	
Particle Coordinates; Quantum Numbers. Calculated and Experimental Values	**Screening Coefficients; Order Numbers**
$n_A(0,0,0)$; $n_B(\underline{\sigma}_1 + \underline{\sigma}_2, 0, 0)$	$s_{1,A1} = s_{A3,A1} = 0.83882$
Phase A:	$s_{A1,1} = s_{A1,A3} = 0.00221203$
$e_1\left(\frac{3}{4}\underline{\sigma}_1, \frac{3}{4}\underline{a}, 0\right)$; $e_2\left(\frac{3}{4}\underline{\sigma}_1, -\frac{3}{4}\underline{a}, 0\right)$	$s_{A1,A2} = 0.25$
Phase B:	$s_{1,2(A)} = \sqrt{\underline{\sigma}_1^2 + \underline{a}^2}/(4\underline{a})$
$e_1\left(\underline{\sigma}_1 + \frac{1}{4}\underline{\sigma}_2, \frac{3}{4}\underline{a}, 0\right)$; $e_2\left(\underline{\sigma}_1 + \frac{1}{4}\underline{\sigma}_2, -\frac{3}{4}\underline{a}, 0\right)$	$s_{1,A3} = (\sqrt{2}/4)/\sqrt{1 + \dfrac{\underline{\sigma}_1}{\sqrt{\underline{\sigma}_1^2 + \underline{a}^2}}}$
$e_{A3}(-\underline{r}_{A3}, 0, 0)$	$s_{1,2(B)} = \sqrt{\underline{\sigma}_2^2 + \underline{a}^2}/(4\underline{a})$
$n_{1(A)} = n_{2(A)} = 2$; $n_{1(B)} = n_{2(B)} = 1$	$Z_{1(A)} = Z_A - 2s_{1,A1} - s_{1,2(A)} - s_{1,A3}$
$n_{A1} = n_{A2} = 1$; $n_{A3} = 2$	$Z_{1M(A)} = Z_A - 2 - s_{1,2(A)} - s_{1,A3}$
$\underline{E}_{b1} = -2.51578$; $\underline{E}_{b2} = -2.11007$	$Z_{A1} = Z_A - s_{A1,A2} - 3s_{A1,1}$
$\underline{\sigma}_1 = 1.34462$; $\underline{\sigma}_2 = 0.0149424$	$Z_{A3} = Z_A - 2s_{1,A1} - 2s_{1,A3}$
$\underline{a} = 0.979054$; $\underline{T}_m = 0.015$	$Z_{1(B)} = Z_B - s_{1,2(B)}$
$\underline{S}_{EiA} = 29.3377$ (Lide, 2003)	$Z_{1M(B)} = Z_B - s_{1,2(B)}$
$\underline{S}_{EiB} = 1$	$\underline{r}_{A3} = 3/(Z_A - 2s_{A3,A1})$
$\underline{D}_0^0 = 0.1495$ (Huber and Hertzberg, 1979)	$Z_{nA} = Z_A - 2 - \dfrac{(\underline{\sigma}_1 + \underline{\sigma}_2)^2}{\left(\underline{\sigma}_1 + \underline{\sigma}_2 + \underline{r}_{A3}\right)^2}$
$\mu_e = 0.192613$ D	$Z_{nB} = Z_B$
$\mu_{eHF} = (0.232, 0.595)$ D (CCCBDB)	

The values of $\underline{E}$, $\underline{E}_{HF}$, and $\underline{E}_{exp}$ are given in Table 2.1. The values of r_0 and r_e are given in Table 2.2. The electric dipole moment is given by the following relation:

$$\mu_e = -5.08315[-2\underline{\sigma}_1 + Z_B(\underline{\sigma}_1 + \underline{\sigma}_2) + \underline{r}_{A3}] \qquad (2.112)$$

The value of μ_e and the domain of the values of μ_{eHF} (CCCBDB) are given in Table 2.9. The calculations are made with the Mathematica 7 program which is given in the Section B.2.8.

2.5.7 The BH Molecule

The structure of this molecule is shown in Figure 2.9B. This molecule contains two electrons, denoted by e_{A3} and e_{A4}, which do not participate to the bond. The geometrical structure of the molecule results from the system (2.65), (2.68), and (2.69), which can be written in normalized form, as follows:

$$
\underline{E}_b = -\frac{Z_{1(A)}^2}{n_{1(A)}^2} - \frac{Z_{2(B)}^2}{n_{2(B)}^2} - \frac{4Z_{1M(B)}}{\sqrt{9\underline{a}^2 + (\underline{\sigma}_1 + 4\underline{\sigma}_2)^2}} - \frac{4Z_{1M(A)}}{\sqrt{9\underline{a}^2 + (\underline{\sigma}_2 + 4\underline{\sigma}_1)^2}}
$$
$$
+ \frac{4}{\sqrt{36\underline{a}^2 + (\underline{\sigma}_1 + \underline{\sigma}_2)^2}} + \frac{Z_{nA}Z_{nB}}{\sigma_1 + \sigma_2}
$$

$$(2.113)$$

$$
\underline{E}_b = \frac{[n_{1(A)} + n_{1(B)}]\sqrt{Z_{1M(A)}}}{(\underline{\sigma}_1^2 + \underline{a}^2)^{3/4}}
$$

$$(2.114)$$

$$
\underline{E}_b = \frac{[n_{1(A)} + n_{1(B)}]\sqrt{Z_{1M(B)}}}{(\underline{\sigma}_2^2 + \underline{a}^2)^{3/4}}
$$

$$(2.115)$$

$$
\underline{E}_b = -\frac{2Z_{1M(A)}}{\sqrt{\underline{\sigma}_1^2 + \underline{a}^2}} - \frac{2Z_{1M(B)}}{\sqrt{\underline{\sigma}_2^2 + \underline{a}^2}} + \frac{1}{2\underline{a}} + \frac{Z_{nA}Z_{nB}}{\underline{\sigma}_1 + \underline{\sigma}_2}
$$

$$(2.116)$$

The system (2.113)–(2.116) has four equations with four unknowns, which are $\underline{E}_b$, $\underline{\sigma}_1$, $\underline{\sigma}_2$, and $\underline{a}$. The expressions of the screening coefficients, order numbers, and quantum numbers are given in Table 2.10. The screening coefficients between the 1s electrons and the e_1, e_2, e_{A3}, and e_{A4} electrons, namely, $s_{1,A1}$, $s_{A1,1}$, $s_{A3,A1}$, and $s_{A1,A3}$, are identical, respectively to s_{31e} and s_{13e}, which have been calculated in Chapter 1 for the boron atom (see Table 1.4). The angle α is a parameter. We perform the calculation for a lot of values of this angle and choose the value for which the total energy is minimum. This value is given in Table 2.10. Using the notations from Figure 2.9B and Eq. (B.14) from Volume I, we have $\underline{r} = \underline{r}_{A3} = \underline{r}_{A4} = (3/Z_{A3})(1 + e^2/2)$, where e is the eccentricity of the valence C_a curves of boron.

Taking into account the symmetry of the system, the total energy results from the following relation (Figure 2.9B):

Table 2.10 Data for BH Molecule	
Particle Coordinates; Quantum Numbers. Calculated and Experimental Values	**Screening Coefficients; Order Numbers**
$n_A(0,0,0)$; $n_B(\underline{\sigma}_1 + \underline{\sigma}_2, 0, 0)$	$s_{1,A1} = s_{A3,A1} = 0.862983$
$e_1\left(\frac{3}{4}\underline{\sigma}_1, \frac{3}{4}\underline{a}, 0\right)$; $e_2\left(\underline{\sigma}_1 + \frac{1}{4}\underline{\sigma}_2, -\frac{3}{4}\underline{a}, 0\right)$	$s_{A1,1} = s_{A1,A3} = 0.00246735$
$e_{A3}(-\underline{d}, 0, \underline{b})$; $e_{A4}(-\underline{d}, 0, -\underline{b})$	$s_{A1,A2} = 0.25$
$n_{1(A)} = n_{2(A)} = 2$; $n_{1(B)} = n_{2(B)} = 1$	$s_{1,A3} = (\sqrt{2}/4)/\sqrt{1 + \dfrac{\underline{\sigma}_1 \cos\alpha}{\sqrt{\underline{\sigma}_1^2 + \underline{a}^2}}}$
$n_{A1} = n_{A2} = 1$; $n_{A3} = n_{A4} = 2$	$s_{A3,A4} = 1/(4\sin\alpha)$
$\underline{E}_b = -3.19537$	$Z_{1(A)} = Z_{2(A)} = Z_A - 2s_{1,A1} - 2s_{1,A3}$
$\underline{\sigma}_1 = 0.888324$; $\underline{\sigma}_2 = 0.261022$	$Z_{1M(A)} = Z_{2M(A)} = Z_A - 2 - 2s_{1,A3}$
$\underline{a} = 0.922599$	$Z_{A1} = Z_A - s_{A1,A2} - 3s_{A1,1}$
$\alpha = 63°$	$Z_{A3} = Z_A - 2s_{1,A1} - s_{1,A3} - s_{A3,A4}$
$\underline{S}_{EiA} = 49.3177$ (Lide, 2003)	$Z_{A4} = Z_{A3}$
$\underline{S}_{EiB} = 1$	$Z_{1(B)} = Z_{2(B)} = Z_B$; $Z_{1M(B)} = Z_{2M(B)} = Z_B$
$\underline{D}_0^0 = 0.251372$ (Huber and Hertzberg, 1979)	$\underline{r} = (2/Z_{A3})/(1 + e^2/2)$
$\mu_e = -1.86638$ D	where $e = 0.97$ for boron
$\mu_{eHF} = (-0.994, -1.814)$ D (CCCBDB)	$\underline{d} = \underline{r}\cos\alpha$; $\underline{b} = \underline{r}\sin\alpha$
	$Z_{nA} = Z_A - 2 - \dfrac{2(\underline{\sigma}_1 + \underline{\sigma}_2)^2(\underline{\sigma}_1 + \underline{\sigma}_2 + \underline{d})}{[(\underline{\sigma}_1 + \underline{\sigma}_2 + \underline{d})^2 + \underline{b}^2]^{3/2}}$
	$Z_{nB} = Z_B$

$$\underline{E} = \underline{E}_1 + \underline{E}_2 + 2\underline{E}_{A3} + 2\underline{E}_{A1} + \underline{U}_{e_1 e_2} + \underline{U}_{e_1 n_B} + \underline{U}_{e_2 n_A} + 2\underline{U}_{e_{A3} e_2}$$

$$+ 2\underline{U}_{e_{A3} n_B} + \underline{U}_{n_A n_B} + \underline{E}_{m1s} = -\frac{Z_{1(A)}^2}{n_{1(A)}^2} - \frac{Z_{2(B)}^2}{n_{2(B)}^2} - 2\frac{Z_{A3}^2}{n_{A3}^2} - 2\frac{Z_{A1}^2}{n_{A1}^2}$$

$$+ \frac{1}{\left[\left(\frac{1}{4}\underline{\sigma}_1 + \frac{1}{4}\underline{\sigma}_2\right)^2 + \left(\frac{3}{2}\underline{a}\right)^2\right]^{1/2}} - \frac{Z_B}{\left[\left(\frac{1}{4}\underline{\sigma}_1 + \underline{\sigma}_2\right)^2 + \left(\frac{3}{4}\underline{a}\right)^2\right]^{1/2}}$$

$$- \frac{Z_A - 2}{\left[\left(\frac{1}{4}\underline{\sigma}_2 + \underline{\sigma}_1\right)^2 + \left(\frac{3}{4}\underline{a}\right)^2\right]^{1/2}} + \frac{2}{\left[\left(\underline{\sigma}_1 + \frac{1}{4}\underline{\sigma}_2 + \underline{d}\right)^2 + \left(\frac{3}{4}\underline{a}\right)^2 + \underline{b}^2\right]^{1/2}}$$

$$- \frac{2Z_B}{[(\underline{\sigma}_1 + \underline{\sigma}_2 + \underline{d})^2 + \underline{b}^2]^{1/2}} + \frac{(Z_A - 2)Z_B}{\underline{\sigma}_1 + \underline{\sigma}_2} + \frac{Z_{A1}^{3/2}}{8n_{A1}^3}$$

$$(2.117)$$

The values of $\underline{E}$, $\underline{E}_{HF}$, and $\underline{E}_{exp}$ are given in Table 2.1. The values of r_0 and r_e are given in Table 2.2. The electric dipole moment is given by the following relation:

$$\mu_e = -5.08315[-2\underline{\sigma}_1 + Z_B(\underline{\sigma}_1 + \underline{\sigma}_2) + 2\underline{d}] \qquad (2.118)$$

The value of μ_e and the domain of the values of μ_{eHF} (CCCBDB) are given in Table 2.10. The Mathematica 7 program for the calculation of the properties of the BH molecule is given in the Section B.2.9.

2.5.8 The CH Molecule

The structure of this molecule is shown in Figure 2.9C. This molecule contains three electrons, denoted by e_{A3}, e_{A4}, and e_{A5}, which do not participate to the bond. The geometrical structure of the molecule results from the system (2.113)–(2.116). All the data necessary in calculations, together with the results of the calculations, are given in Table 2.11. Using the notations from Figure 2.9C and Eq. (B.14) from Volume I, we have $\underline{r} \cong (\underline{r}_{A3} + \underline{r}_{A4} + \underline{r}_{A5})/3 = (2/3)(1/Z_{A3} + 1/Z_{A4} + 1/Z_{A5})(1 + e^2/2)$, where e is the eccentricity of the valence C_a curves of carbon.

The normalized value of the total energy is calculated with the aid of Eq. (2.77), which can be written as follows:

$$
\begin{aligned}
\underline{E} &= \underline{E}_1 + \underline{E}_2 + \underline{E}_{A3} + \underline{E}_{A4} + \underline{E}_{A5} + 2\underline{E}_{A1} + \underline{U}_{e_1 e_2} + \underline{U}_{e_1 n_B} \\
&\quad + \underline{U}_{e_2 n_A} + \underline{U}_{e_{A3} e_2} + \underline{U}_{e_{A4} e_2} + \underline{U}_{e_{A5} e_2} + 3\underline{U}_{e_{A3} n_B} + \underline{U}_{n_A n_B} + \underline{E}_{m1s} \\
&= -\frac{Z_{1(A)}^2}{n_{1(A)}^2} - \frac{Z_{2(B)}^2}{n_{2(B)}^2} - \frac{Z_{A3}^2}{n_{A3}^2} - \frac{Z_{A4}^2}{n_{A4}^2} - \frac{Z_{A5}^2}{n_{A5}^2} - 2\frac{Z_{A1}^2}{n_{A1}^2} \\
&\quad + \frac{1}{\left[\left(\frac{1}{4}\underline{\sigma}_1 + \frac{1}{4}\underline{\sigma}_2\right)^2 + \left(\frac{3}{2}\underline{a}\right)^2\right]^{1/2}} - \frac{Z_B}{\left[\left(\frac{1}{4}\underline{\sigma}_1 + \underline{\sigma}_2\right)^2 + \left(\frac{3}{4}\underline{a}\right)^2\right]^{1/2}} \\
&\quad - \frac{Z_A - 2}{\left[\left(\frac{1}{4}\underline{\sigma}_2 + \underline{\sigma}_1\right)^2 + \left(\frac{3}{4}\underline{a}\right)^2\right]^{1/2}} + \frac{1}{\left[\left(\underline{\sigma}_1 + \frac{1}{4}\underline{\sigma}_2 + \underline{d}\right)^2 + \left(\frac{3}{4}\underline{a}\right)^2 + \underline{b}^2\right]^{1/2}} \quad (2.119) \\
&\quad + \frac{1}{\left[\left(\underline{\sigma}_1 + \frac{1}{4}\underline{\sigma}_2 + \underline{d}\right)^2 + \left(\frac{3}{4}\underline{a} + \frac{\sqrt{3}}{2}\underline{b}\right)^2 + \left(\frac{1}{2}\underline{b}\right)^2\right]^{1/2}} \\
&\quad + \frac{1}{\left[\left(\underline{\sigma}_1 + \frac{1}{4}\underline{\sigma}_2 + \underline{d}\right)^2 + \left(\frac{3}{4}\underline{a} - \frac{\sqrt{3}}{2}\underline{b}\right)^2 + \left(\frac{1}{2}\underline{b}\right)^2\right]^{1/2}} \\
&\quad - \frac{3Z_B}{[(\underline{\sigma}_1 + \underline{\sigma}_2 + \underline{d})^2 + \underline{b}^2]^{1/2}} + \frac{(Z_A - 2)Z_B}{\underline{\sigma}_1 + \underline{\sigma}_2} + \frac{Z_{A1}^{3/2}}{8n_{A1}^3}
\end{aligned}
$$

Table 2.11 Data for CH Molecule

Particle Coordinates; Quantum Numbers. Calculated and Experimental Values	Screening Coefficients; order Numbers
$n_A(0,0,0)$; $n_B(\underline{\sigma}_1 + \underline{\sigma}_2, 0, 0)$	$s_{1,A1} = 0.85505$
$e_1\left(\frac{3}{4}\underline{\sigma}_1, \frac{3}{4}\underline{a}, 0\right)$; $e_2\left(\underline{\sigma}_1 + \frac{1}{4}\underline{\sigma}_2, -\frac{3}{4}\underline{a}, 0\right)$	$s_{A1,1} = 0.00269287$
$e_{A3}(-\underline{d}, 0, \underline{b})$; $e_{A4}\left(-\underline{d}, \frac{\sqrt{3}}{2}\underline{b}, -\frac{1}{2}\underline{b}\right)$; $e_{A5}\left(-\underline{d}, -\frac{\sqrt{3}}{2}\underline{b}, -\frac{1}{2}\underline{b}\right)$	$s_{1,A1} = s_{A3,A1} = s_{A4,A1} = s_{A5,A1}$
$n_{1(A)} = n_{2(A)} = 2$; $n_{1(B)} = n_{2(B)} = 1$.	$s_{A1,1} = s_{A1,A3} = s_{A1,A4} = s_{A1,A5}$
$n_{A1} = n_{A2} = 1$; $n_{A3} = n_{A4} = n_{A5} = 2$	$s_{A1,A2} = 0.25$
$\underline{E}_b = -4.29182$	$s_{1,A3} = (\sqrt{2}/4)\Big/ \sqrt{1 + \dfrac{\underline{\sigma}_1 \cos\alpha}{\sqrt{\underline{\sigma}_1^2 + \underline{a}^2}}}$
$\underline{\sigma}_1 = 0.840475$; $\underline{\sigma}_2 = 0.175936$	$s_{1,A4} = (\sqrt{2}/4)\Big/ \sqrt{1 + \dfrac{\underline{\sigma}_1 \cos\alpha - \frac{\sqrt{3}}{2}\underline{a}\sin\alpha}{\sqrt{\underline{\sigma}_1^2 + \underline{a}^2}}}$
$\underline{a} = 0.767725$	$s_{1,A5} = (\sqrt{2}/4)\Big/ \sqrt{1 + \dfrac{\underline{\sigma}_1 \cos\alpha + \frac{\sqrt{3}}{2}\underline{a}\sin\alpha}{\sqrt{\underline{\sigma}_1^2 + \underline{a}^2}}}$
$\alpha = 67°$	$s_{A3,A4} = 1/(2\sqrt{3}\sin\alpha)$
$\underline{S}_{EiA} = 75.7133$ (Lide, 2003)	$Z_{1(A)} = Z_A - 2s_{1,A1} - s_{1,A3} - s_{1,A4} - s_{1,A5}$
$\underline{S}_{EiB} = 1$	$Z_{1M(A)} = Z_A - 2 - s_{1,A3} - s_{1,A4} - s_{1,A5}$
$\underline{D}_0^0 = 0.254679$ (Huber and Hertzberg, 1979)	$Z_{1(A)} = Z_{2(A)}$; $Z_{1M(A)} = Z_{2M(A)}$
$\mu_e = -1.91195$ D	$Z_{A1} = Z_A - s_{A1,A2} - 4s_{A1,1}$
$\mu_{eHF} = (-1.069, -1.666)$ D (CCCBDB)	$Z_{A3} = Z_A - 2s_{1,A1} - s_{1,A3} - 2s_{A3,A4}$
	$Z_{A4} = Z_A - 2s_{1,A1} - s_{1,A4} - 2s_{A3,A4}$
	$Z_{A5} = Z_A - 2s_{1,A1} - s_{1,A5} - 2s_{A3,A4}$
	$Z_{1(B)} = Z_{2(B)} = Z_B$
	$Z_{1M(B)} = Z_{2M(B)} = Z_B$
	$e = 0.98$ for carbon
	$\underline{r} = \left(\dfrac{1}{Z_{A3}} + \dfrac{1}{Z_{A4}} + \dfrac{1}{Z_{A5}}\right)\frac{2}{3}\left(1 + \frac{e^2}{2}\right)$
	$\underline{d} = \underline{r}\cos\alpha$; $\underline{b} = \underline{r}\sin\alpha$
	$Z_{nA} = Z_A - 2 - \dfrac{3(\underline{\sigma}_1 + \underline{\sigma}_2)^2(\underline{\sigma}_1 + \underline{\sigma}_2 + \underline{d})}{[(\underline{\sigma}_1 + \underline{\sigma}_2 + \underline{d})^2 + \underline{b}^2]^{3/2}}$
	$Z_{nB} = Z_B$

The values of $\underline{E}$, $\underline{E}_{HF}$, and $\underline{E}_{exp}$ are given in Table 2.1. The values of r_0 and r_e are given in Table 2.2. The electric dipole moment is given by the following relation:

$$\mu_e = -5.08315[-2\underline{\sigma}_1 + Z_B(\underline{\sigma}_1 + \underline{\sigma}_2) + 3\underline{d}] \qquad (2.120)$$

The value of α, corresponding to the minimum value of $\underline{E}$, together with the value of μ_e and the domain of the values of μ_{eHF} (CCCBDB) are given in Table 2.11. The Mathematica 7 program for the calculation of the properties of the BH molecule is given in Section B.2.10.

CHAPTER 3

Modeling Properties of Harmonics Generated by Relativistic Interactions Between Very Intense Electromagnetic Beams, Electrons, and Atoms

Abstract

The connection between the Klein–Gordon and relativistic Hamilton–Jacobi equations, proved in Chapter 2 from Volume I for the system electron–electromagnetic field, justifies the accuracy of the classical approaches in the new field of interactions between very intense laser beams and electrons or atoms. In the same time, the periodicity property, also proved in Chapter 2 from Volume I, leads to accurate classical models for these interactions. We show that the relations which result from the periodicity property lead to accurate calculations of the angular and spectral distributions of the radiations emitted at the interactions between very intense laser beams and electron plasmas, relativistic electron beams and atomic gases. Our theoretical results explain the main properties of these interactions and they are in good agreement with numerous experimental data from the literature.

Keywords: Very intense laser beams; electron plasma; relativistic electron beam; Thomson scattering; head-on collisions; hard radiations; high harmonic spectrum; angular distributions; spectral distributions; polarization effects; laser–atom interaction

3.1 GENERAL CONSIDERATIONS

In this chapter, we present a review of the applications presented in our previous papers (Popa, 2008b, 2009b, 2011b, 2012) for systems composed of electromagnetic fields and particles. In numerous papers from the literature, including our papers (Popa et al., 1997; Popa, 2003b, 2004, 2007), it is shown that classical models of interaction between an electromagnetic field and electrons are in agreement with the experimental data for a large range of values of the electromagnetic beam intensity, going from medium values to the very high values achievable in modern systems employing very intense lasers. In this chapter, we show that the periodicity property proved in Chapter 2 from Volume I leads to exact solutions for spectral and angular distributions of the intensities of the radiations resulted at the interaction between electron plasmas or relativistic electron beams, and very intense laser beams. In the same time, we show that a classical treatment of the ionization domain in the frame of the Corkum's model (Corkum, 1993) is justified in interactions between laser beams and atoms. It follows that the energy of the electron that returns in the vicinity of the atom, in the frame of the Corkum's model, is accurately calculated. Introducing this value in the formula of Lewenstein et al. (1994) that gives the probability of emitting harmonic radiations, we obtain an accurate method to calculate the shape of the emitted radiation spectrum. This theoretical result is in good agreement with numerous experimental data reported in the literature for many atoms and ions and for a large range of laser beam intensities.

Throughout this chapter we use the notations introduced in Chapter 2 from Volume I.

3.2 RADIATIONS GENERATED AT THE INTERACTIONS BETWEEN VERY INTENSE LASER BEAMS AND ELECTRON PLASMAS

We apply the algorithm described in Section 2.4 from Volume I, when the initial data are a_1, a_2, β_{xi}, β_{yi}, β_{zi}, η_i, θ, ϕ, and j. With the aid of relations from Sections 2.3 and 2.4 from Volume I, we calculate f_1, f_2, γ, f_3, g_1, g_2, g_3, f_{1sj}, f_{2sj}, f_{3sj}, f_{1cj}, f_{2cj}, f_{3cj}, the normalized average intensity, $\underline{I}_{av}$, and the normalized average intensity of the jth harmonic, $\underline{I}_j$, generated at the interaction between very intense laser beam and electron. The calculations are made with the aid of Program B.3.1, given in Appendix B. We present below typical angular and spectral distributions of the scattered beam intensities.

3.2.1 Typical Angular Distributions

In Figure 3.1, we show the angular distributions of $\underline{I}_{av}$, normalized to the maximum values, versus θ in the plane xy, at the interaction between a circular polarized field, which propagates in the oz direction, and electrons, in two cases: (i) the electrons move in the opposite

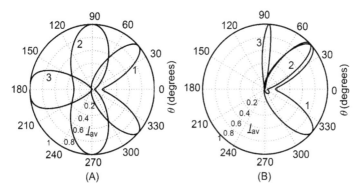

Figure 3.1 Polar plots of $\underline{I}_{av}$, normalized to maximum values, as functions of θ, at the interaction between a circular polarized field, having $a_1 = a_2 = 2$, which propagates in the oz direction, and electrons, when $\phi = 0$ and $\eta_i = 0$, in two cases. (A) The electrons move in the opposite direction, namely $\beta_{xi} = \beta_{yi} = 0$. Calculations are made for three cases, when $\gamma_i = 1$ and $\beta_{zi} = 0$ (curve 1), $\gamma_i = 1.189$ and $\beta_{zi} = -0.5410$ (curve 2), and $\gamma_i = 3$ and $\beta_{zi} = -0.9428$ (curve 3). (B) Electrons move in the ox direction, namely $\beta_{yi} = \beta_{zi} = 0$. Calculations are made for three cases, when $\gamma_i = 1$ and $\beta_{xi} = 0$ (curve 1), $\gamma_i = 1.1$ and $\beta_{xi} = 0.4166$ (curve 2), and $\gamma_i = 10$ and $\beta_{xi} = 0.9950$ (curve 3).

direction and (ii) the electrons move in the *ox* direction. The data from our calculations are given in the caption of the figure.

Figure 3.1A shows that when the electron moves in the opposite direction, compared to the direction of the laser beam, the scattering emission lobes are symmetrical and, when the initial electron velocity is small, the radiation is emitted in the *oz* direction. When the velocity increases, the angle between the axis of the emission lobe and the *oz* axis increases. When the initial electron velocity is sufficiently big, this electron emits radiation in the backward direction with respect to the propagation direction of the incident wave. For very high values of $|\beta_{zi}|$, there is only one backscattered emission lobe. In Figure 3.1B, when the electron moves in the direction perpendicular to the direction of the incident wave, and $|\beta_{xi}|$ increases, the emission lobes are asymmetrical, one lobe is diminished and the bigger emission lobe is shifted toward the *ox* direction.

A comparison between Figure 3.1 and figures 1 and 2 from Goreslavskii et al. (1999) shows that the general properties of the scattered beam are the same in both cases, in spite of the fact that the treatment of Goreslavskii is completely different of our treatment.

In the same manner, it is easy to show that in electron plasma, in which the directions of the moving of the electrons are random, the lobes of the emission of the radiation, for any two electrons, are different. It follows that in the case of electron plasma, the overall angular distribution of $\underline{I}_{av}$ is broader than the theoretical prediction when the electron has a very small velocity, as it is shown by Popa (2011b).

3.2.2 Typical Spectral Distributions

We study first the spectral distributions of the scattered radiations in interactions between very intense laser beams and electron plasmas created by above threshold ionization of an atomic gas. In this case, the kinetic energies of the electrons are of the order of 50 eV (Agostini and DiMauro, 2008), resulting that their normalized velocities are of the order of 0.014. We consider an arbitrary case, corresponding to $|\overline{\beta}_i| = 0.014$, for different directions of the velocities in the plane xz. In addition, we consider the case when the velocities of the electrons are zero. It is easy to show that the spectral distributions corresponding to all these velocities lie between the curves that correspond to the forward and backward directions of the initial electron velocities, and these curves are close to the central curve that corresponds to the case when the electron velocities are negligible. It follows that we can consider, in this case, that the effect of the initial velocities on the spectrum of the scattered radiations is negligible.

In Figure 3.2, we plot the harmonics spectrum in the general case, when the incident electromagnetic field is elliptic polarized for arbitrary values of the input parameters and for different values of a_1. Two

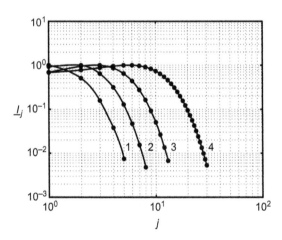

Figure 3.2 Spectrum of the scattered Thomson radiation, where the intensities $\underline{I}_j$ are normalized to their maximum values, in the case of the interaction between an elliptic polarized field, having $a_2 = 5$, $\theta = 15°$, $\phi = 0°$, $\eta_i = 35°$, for negligible value of the initial electron velocity. The curves 1, 2, 3, and 4 correspond, respectively, to $a_1 = 2$, $a_1 = 4$, $a_1 = 6$, and $a_1 = 8$. The continuous curves interpolate the discrete values $\underline{I}_j$ denoted by points.

properties are revealed by this figure. First, the spectrum has a specific shape, characterized by a slowly increasing portion for low order harmonics, followed by a maximum, and a sharper decreasing portion for the highest orders of the harmonics. The second property is that the spectrum has a maximum shifting to shorter wavelengths, as a_1 increases.

3.3 HARD RADIATIONS GENERATED AT THE HEAD-ON COLLISION BETWEEN VERY INTENSE LASER BEAM AND RELATIVISTIC ELECTRON BEAM

We apply the algorithm described in Section 2.4 from Volume I, when the analysis is made in the system S' and the initial data are a_1, a_2, γ_0, η_i, θ', ϕ', ω_L, and j. With the aid of relations from Section 2.5 from Volume I, we calculate $f'_1, f'_2, \gamma', f'_3, g'_1, g'_2, g'_3, F'_1, F'_2, h'_1, h'_2, h'_3, f'_{1sj}$, $f'_{1cj}, f'_{2sj}, f'_{2cj}, f'_{3sj}, f'_{3cj}, \underline{I}'_j, \underline{I}_j$, and W_j. The calculations are made with the aid of Program B.3.2, given in Appendix B. We present below typical angular and spectral distributions of the scattered beam intensities.

3.3.1 Typical Angular Distributions

We consider the head-on collision between a linearly polarized laser beam, having $a_1 = 0.35$, $a_2 = 0$, $\lambda_L = 10.6\ \mu m$, and $\omega_L = 1.777 \times 10^{14}$ rad/s, with relativistic electrons having the energy $E_i = 60$ MeV that corresponds to $\gamma_0 = 117.4$. The values of the intensities of the fundamental and second harmonic component, normalized to $\varepsilon_0 c K'^2$, denoted by $\underline{I}_1$ and $\underline{I}_2$, and calculated with the aid of Program B.3.2, are shown in Table 3.1, as functions of θ', when θ'

Table 3.1 Variations of $\underline{I}_1$, $\underline{I}_2$, θ, and α, as Functions of θ', for $E_i = 60$ MeV, $\gamma_0 = 117.4$, $a_1 = 0.35$, $a_2 = 0$, $\omega_L = 1.777 \times 10^{14}$ rad/s, $\eta_i = 0$, and $\phi = 0$

θ' (deg)	90	100	110	120	130	140
θ (rad)	3.1331	3.1345	3.1356	3.1367	3.1376	3.1385
α (mrad)	8.5180	7.1475	5.9644	4.9179	3.9720	3.1003
$\underline{I}_1$	52.67	811.1	2794	6102	10,429	15,186
$\underline{I}_2$	355.5	955.6	1779	2537	2894	2674
θ' (deg)	150	160	170	175	180	
θ (rad)	3.1393	3.1401	3.1409	3.1412	3.1416	
α (mrad)	2.2824	1.5020	0.7452	0.3719	0	
$\underline{I}_1$	19,694	23,352	25,720	26,332	26,538	
$\underline{I}_2$	1956	1037	286.9	73.51	0	

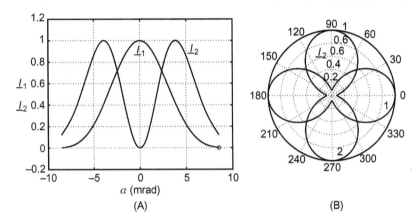

Figure 3.3 Angular variations of $\underline{I}_1$ and $\underline{I}_2$, normalized to their maximum values for $\gamma_0 = 117.4$, $a_1 = 0.35$, $a_2 = 0$, $\omega_L = 1.777 \times 10^{14}$ rad/s, and $\eta_i = 0$. (A) Variations of $\underline{I}_1$ and $\underline{I}_2$ with α, for $\phi = 0$ and (B) variations of $\underline{I}_2$ as function of ϕ, for $\theta' = 160°$, when the electromagnetic field is linearly polarized in the directions ox (curve 1) and oy (curve 2).

takes values between $90°$ and $180°$. With the aid of Eq. (2.127) from Volume I, we calculate the values of θ as functions of θ'. In this table, we give also the values of the angle α, given by the relation

$$\alpha = \pi - \theta \tag{3.1}$$

Taking into account the values shown in Table 3.1, in Figure 3.3A we plot the angular variations of $\underline{I}_1$ and $\underline{I}_2$, normalized to their maximum values, as functions of α, in the plane of the polarization, corresponding to $\phi = 0$. These curves are symmetrical with respect to the direction corresponding to the angle $\alpha = 0$.

For the same system, in Figure 3.3B we plot the polar variations of the intensity of the second harmonics, $\underline{I}_2$, normalized to its maximum value, as function of ϕ, for constant θ', when the electromagnetic field is linearly polarized in the directions ox and oy.

Our calculations presented in Figure 3.3 are in good agreement with the experimental results presented in figures 3 and 5 of Babzien et al. (2006), which are obtained for the same initial data, as those considered in our calculations. On the other hand, in virtue of Eq. (2.132) from Volume I for $\theta' = \pi$, the energies corresponding to the fundamental and second harmonic backscattered radiations are given by $W_{b1} = \omega_L \gamma_0^2 (1 + |\overline{\beta}_0|)^2 \hbar$ and $W_{b2} = 2\omega_L \gamma_0^2 (1 + |\overline{\beta}_0|)^2 \hbar$ and lead,

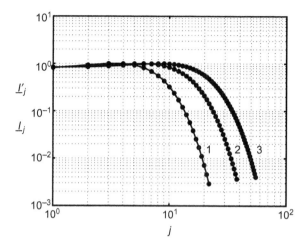

Figure 3.4 Typical spectral distributions for the backscattered Thomson radiation, namely, the variations of I'_j and I_j, in the inertial systems S' and S, normalized to their maximum values, for $a_2 = 4$, $\gamma_0 = 156.5$, $\omega_L = 2.355 \times 10^{15}$ rad/s, $\theta = \theta' = 180°$ and $\eta_i = 25°$, for $a_1 = 3$ (curve 1), 6 (curve 2), and 9 (curve 3). The spectra I'_j and I_j are identical. The continuous curves interpolate the discrete values of I'_j and I_j denoted by points.

respectively, to the values $W_{b1} = 6.45$ keV and $W_{b2} = 12.90$ keV. These values are in very good agreement with the experimental values, which are reported by Babzien et al. (2006).

3.3.2 Typical Spectral Distributions

With the aid of Program B.3.2 from Appendix B, we calculate the spectra of the backscattered beam in the inertial systems S' and S, when I'_j and I_j are normalized to their maximum values. These spectra are presented in Figure 3.4. The spectra are calculated in an arbitrary case, when the incident field is elliptic polarized, and the initial energy of the electron is $E_i = 80$ MeV, resulting, in accord with relation $E_i = \gamma_0 mc^2$, that $\gamma_0 = 156.5$. Our calculations are made for $\omega_L = 2.355 \times 10^{15}$ rad/s (for a Ti:Sapphire laser), $\theta = \theta' = 180°$ and $\eta_i = 25°$, for three values of a_1, equal to 3, 6, and 9. The analysis of the relation (2.116) from Volume I shows that the two spectra are identical.

The spectra from Figure 3.4 have the same shape as the spectra from Figure 3.2 and their maxima are shifting to shorter wavelengths, as a_1 increases.

3.4 EFFECTS IN COLLISIONS AT ARBITRARY ANGLES BETWEEN VERY INTENSE σ_L OR π_L POLARIZED LASER BEAMS, AND RELATIVISTIC ELECTRON BEAMS

We apply the algorithm described in Section 2.4 from Volume I, when the analysis of the collision, at arbitrary angle, between very intense laser beam and relativistic laser beam is made in the inertial system S', in which the electron velocity is zero, and the initial data are θ_L, a, γ_0, η_i, θ', ϕ', ω_L, and j. With the aid of relations from Sections 2.5 and 2.6 from Volume I, we calculate f'_1, f'_2, γ', f'_3, g'_1, g'_2, g'_3, F'_1, F'_2, h'_1, h'_2, h'_3, f'_{1sj}, f'_{1cj}, f'_{2sj}, f'_{2cj}, f'_{3sj}, f'_{3cj}, $\underline{I}'_j$, I_j, ω_j, and W_j for σ_L and π_L polarizations. The calculations are made with the aid of programs slightly different of Program B.3.2, given in Appendix B, as it is shown in Section 2.6.3 from Volume I. We present below typical angular distributions of the scattered beam intensities and applications of the polarization effects, resulted from the theory presented in Section 2.6 from Volume I. The spectral distributions are similar to those presented in Section 3.3, in the case of the head-on collision.

3.4.1 Typical Angular Distributions

Since the θ_L angle between the directions of the laser and electron beams is allowed to have arbitrary values, the widely analyzed 180° and 90° geometries, in which the two beams collide, respectively, head on and perpendicularly, are particular cases. We consider four experiments, corresponding to 180° and 90° geometries, for which we calculate the angular distributions of the scattered radiations. These distributions are presented in Figure 3.5. The calculations are made for $\theta' = 180°$ and for the values of θ_L, a, γ_0, η_i, ϕ', and ω_L listed in caption of the figure. The calculations are made as in the case of Figure 3.3A. We calculate the functions $\underline{I}_1 = \underline{I}_1(\theta')$ and $\alpha = \alpha(\theta')$, corresponding to $j = 1$, where α is given by Eq. (3.1), and obtain the variation of I_1 as function of α.

The curves 1, 2, 3, and 4 from Figure 3.5 are calculated for data identical to those from the experiments presented by Babzien et al. (2006), Pogorelsky et al. (2000), Anderson et al. (2004), and Kim et al. (1994). The values of the divergence angles of the scattered beams, which result from Figure 3.5, are in agreement with the experimental values presented in these references.

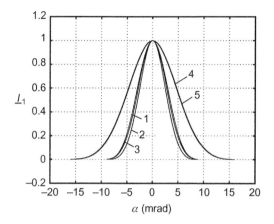

Figure 3.5 Variations of $\underline{I}_1$, normalized to their maximum values, with α. Curve 1 is for the 180° geometry, $\theta_L = 0°$, $\gamma_0 = 117.4$, $a = 0.35$, $\eta_i = 0°$, $\theta' = 180°$, $\phi' = 0°$, and $\omega_L = 1.777 \times 10^{14}$ rad/s; curve 2 is for the 180° geometry, $\theta_L = 0°$, $\gamma_0 = 117.4$, $a = 0.053$, $\eta_i = 0°$, $\theta' = 180°$, $\phi' = 0°$, and $\omega_L = 1.777 \times 10^{14}$ rad/s; curve 3 is for the 180° geometry, $\theta_L = 0°$, $\gamma_0 = 111.5$, $a = 0.2012$, $\eta_i = 0°$, $\theta' = 180°$, $\phi' = 0°$, and $\omega_L = 2.355 \times 10^{15}$ rad/s; curve 4 is for the 90° geometry, σ_L polarization, $\theta_L = 90°$, $\gamma_0 = 62.62$, $a = 0.08369$, $\eta_i = 0°$, $\theta' = 180°$, $\phi' = 0°$, and $\omega_L = 2.355 \times 10^{15}$ rad/s; curve 5 is for the same parameters as for curve 4, but the π_L polarization is used instead.

3.4.2 Polarization Effects

As the angle θ_L increases, the system will pass periodically through the two configurations of interaction having 180° or 90° geometries. The value $\theta_L = 0$ corresponds to the 180° geometry. In this case, the σ_L and π_L polarizations are identical, because they correspond, in virtue of the relations from Section 2.6 from Volume I, to the same incident electromagnetic field. The values $\theta_L = \pm 90°$ correspond to the 90° geometry. For these θ_L values, the σ_L and π_L polarizations behave differently, because, in this case, the incident electromagnetic fields are different.

In Figure 3.6A, we show the variation of the quanta energy of the scattered radiation, for $\theta' = \theta = 180°$, obtained with the aid of relation (2.191) from Volume I, as function of θ_L. The energy is normalized to $j\omega_L\gamma_0^2(1+|\overline{\beta}_0|)^2\hbar$. This figure shows that the energy corresponding to the 90° geometry is half the energy corresponding to the 180° geometry, for same values of ω_L and γ_0.

From the theory presented in Section 2.6.4 from Volume I, it follows that the components of the scattered electromagnetic field emitted in a certain direction are rigorously determined. It follows that for a point on the curve shown in Figure 3.6A, which corresponds a certain value of θ_L and to a certain geometry of the collision between the laser

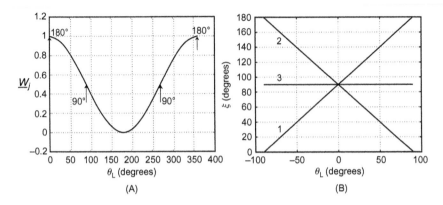

Figure 3.6 (A) Variation of the quanta energy of the scattered radiation normalized to $j\omega_L\gamma_0^2(1+|\vec{\beta}_0|)^2\hbar$ as function of θ_L, for $\theta' = \theta = 180°$ and $\gamma_0 = 100$. (B) Variations of the angle ξ, which corresponds to the polarization of the electric field of the scattered radiation, as function of θ_L, for $a = 4$, $\gamma_0 = 100$, $\theta' = \theta = 180°$, $\eta_i = 20°$, and $\omega_L = 2.355 \times 10^{15}$ rad/s. Curve 1 corresponds to the σ_L polarization for odd harmonics of the scattered radiation, curve 2 corresponds to the σ_L polarization for even harmonics, while curve 3 corresponds for the π_L polarization, for both the even and odd harmonics. The σ polarization of the scattered field corresponds to $\xi = 0°$, while the π polarization corresponds to $\xi = \pm 90°$.

and electron beams, we can evaluate exactly the polarization of the scattered beam.

The polarization which corresponds to a given interaction geometry results from Figure 3.6B, in which we represent the variations of the angle ξ with θ_L, in the cases of the the σ_L and π_L polarizations of the incident field, for odd and even harmonics of the scattered radiation. The angle ξ corresponds to the polarization of the electric field of the scattered radiation.

From Figure 3.6A and B, we can see that the energy and the polarization of the scattered radiation can be accurately adjusted by the variation of the angle θ_L.

3.5 CALCULATION OF THE HARMONIC SPECTRUM OF THE RADIATIONS GENERATED AT THE INTERACTION BETWEEN VERY INTENSE LASER BEAMS AND ATOMS

The calculation of the harmonic spectrum is based on two reliable results that are in agreement with numerous experimental data from the literature. The first is the classical model of the ionization domain (Corkum, 1993). The second is the accurate rate of harmonic emission, calculated in the frame of the model from Lewenstein et al. (1994).

In Section 2.7 from Volume I, we have shown that the electron leaves the atom when the phase of the field is η_0, and it returns in the vicinity of the atom when the phase of the field is η_1. It follows that, in virtue of Eq. (2.203) from Volume I, for a given value of η_0, there can be only one value for η_1, and we have $\eta_1 = \eta_1(\eta_0)$. With the aid of this relation, we will show below that all the quantities which are used in our analysis are functions of only one variable, which is η_0.

The probability of tunneling of the electron, when it leaves the atom, normalized to its maximum value, is given by Eq. (2.211) from Valume I, that is

$$\underline{W} = \frac{1}{|\cos\eta_0|^{2n*-m}m - 1} \exp\left[-\left(\frac{1}{|\cos\eta_0|} - 1\right)\frac{4(2m)^{1/2}I_p^{3/2}}{3eE_M\hbar}\right]$$

The rate of emission of the radiation obtained in the transition from the state in which the atom is ionized and the electron is in its vicinity and has the energy E, to the state in which the electron is recombined with the atom, is proportional to $r = \overline{d} \cdot \overline{d}^*$, where $\overline{d}$ is the dipole matrix element corresponding to the electric dipole transition. This rate is calculated with relation (24) from Lewenstein et al. (1994). Writing this relation in the International System, we obtain

$$r = K(2I_p)^{5/2}(ea_0)^2 \frac{\overline{p}^2/m}{(\overline{p}^2/m + 2I_p)^6} \tag{3.2}$$

where

$$K = \frac{2^7}{\pi^2}\left(\frac{e^2}{4\pi\varepsilon_0 a_0}\right)^{5/2} \tag{3.3}$$

and $\overline{p}$ is the kinetic momentum of the electron in the vicinity of the nucleus. In our case $\overline{p} = \overline{p}_1$, where $\overline{p}_1$ corresponds to $\eta = \eta_1$, in agreement with Eq. (2.209) from Volume I. Since $\eta_1 = \eta_1(\eta_0)$, it follows that r is function of only one variable, which is η_0, and have $r = r(\eta_0)$.

The overall rate of emission of the radiation is the product of two probabilities. The first is $\underline{W}$, the probability of tunneling of the electron. From the above equation, it follows that $\underline{W}$ is also a function of only the phase η_0, and we can write $\underline{W} = \underline{W}(\eta_0)$. The second is the rate of the generation of the electromagnetic field, r, when the electron recombines with the atom. In virtue of Eqs (2.209) and (2.211) from

Volume I, and (3.2), we obtain the overall rate of generation of the electromagnetic field, denoted by R:

$$R = \underline{W}r = K\underline{W}(2I_p)^{-5/2}(ea_0)^2 \frac{f}{(f+1)^6} \tag{3.4}$$

where

$$f = \gamma_1(\sin \eta_1 - \sin \eta_0)^2 \left[1 + \frac{a^2}{4}(\sin \eta_1 - \sin \eta_0)^2\right] \tag{3.5}$$

and

$$\gamma_1 = \frac{2U_p}{I_p} \tag{3.6}$$

Taking into account that the relations $\eta_1 = \eta_1(\eta_0)$, Eqs (3.4) and (3.5), it follows that R is a function of only the phase η_0 and has $R = R(\eta_0)$. In the same time, from Eqs (2.203) and (2.207) from Volume I, it follows that the energy of the electromagnetic field is also a function of only the phase η_0 and we have $E_{em} = E_{em}(\eta_0)$. From $R = R(\eta_0)$ and $E_{em} = E_{em}(\eta_0)$, one obtain the curve $R = R(E_{em})$.

The effect represented by the rate r dominates the effect represented by $\underline{W}$. If we neglect the latter, namely if we set $\underline{W} = 1$, then the shape of the spectrum remains roughly the same.

Taking into account the significance of E_{em}, we have

$$E_{em} = nE_q \quad \text{with} \quad E_q = \omega_L \hbar \tag{3.7}$$

where E_q is quanta energy of the incident field and n is the order of the harmonic.

It follows that the corresponding cutoff order of harmonics, denoted by n_c, is the integer part of E_c/E_q, namely

$$n_c = \frac{E_c}{E_q} \tag{3.8}$$

where E_c is given in Eq. (2.210) from Volume I.

In Figure 3.7, we present typical theoretical spectra of the radiations emitted at the interaction of very intense laser beams with atoms, namely the variations $\underline{R} = \underline{R}(E_{em})$, where $\underline{R} = R/(ea_0)^2$, and $\underline{R} = \underline{R}(n)$,

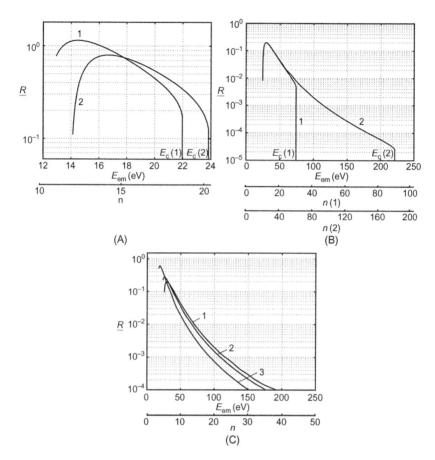

Figure 3.7 Typical harmonic spectra, namely the variations of the harmonic generation rate $\underline{R}$ with quanta energy E_{em} and harmonic order number n, corresponding to different values of I_L. The continuous curves interpolate the discrete values of $\underline{R}$. The data corresponding to each curve are given in Table 3.2.

for three domains of values of the I_L intensities: (i) for I_L of the order of 10^{13} W/cm^2 in Figure 3.7(A), (ii) for I_L of the order of 10^{15} W/cm^2 in Figure 3.7(B), and (iii) for I_L of the order of 10^{17} W/cm^2 in Figure 3.7C. The continuous curves in Figure 3.7 interpolate the discrete values of $\underline{R}$ that result from Eq. (3.7).

The experimental conditions for the curves shown in Figure 3.7 were chosen to be the same as those used in experiments reported in the literature. In this respect, curves 1 and 2 in Figure 3.7A are compared to the experimental curves 2 and 3 from Ferray et al. (1988), curves 1 and 2 from Figure 3.7B are compared, respectively, with the curves represented by thick line and open circles in figure 3 from

Table 3.2 Data for the Harmonic Spectra Presented in Figure 3.7

Figure 3.7	A	A	B	B	C	C	C
Curve	1	2	1	2	1	2	3
Atom	Xe	Kr	He	He	He	Ne	Ar
I_p (eV)	12.130	13.999	24.6	24.6	24.6	21.564	15.760
I_L (W/cm^2)	3×10^{13}	3×10^{13}	6×10^{14}	6×10^{14}	2×10^{17}	4×10^{17}	3×10^{17}
λ_L (μm)	1.053	1.053	0.527	1.053	0.248	0.248	0.248
a	0.00493	0.00493	0.0110	0.0221	0.0948	0.1341	0.1161
U_p (eV)	3.106	3.106	15.56	62.12	1148.6	2297.3	1723.0
E_c (eV)	21.98	23.85	73.93	221.5	3665.8	7304	5477.6
n_c	18	20	31	188	733	1461	1095
E_q (eV)	1.177	1.177	2.353	1.177	5.00	5.00	5.00
γ_1	0.5122	0.4438	1.265	5.051	93.39	213.06	218.7

Krause et al. (1992). Curves 1, 2, and 3 from Figure 3.7C are compared, respectively, to the curves corresponding to He, Ne, and Ar in figure 2 from Sarukura et al. (1991). These comparisons show a good agreement between the theoretical curves presented (Figure 3.7) and the experimental curves reported in the above references.

The values on the ordinate axes of the spectra in the theoretical and experimental figures differ by a scaling factor since R is proportional to the experimental rate of generation of the harmonics, and computing the proportionality factor, although possible, is irrelevant to the topic of this paper, only the shape of the spectra is important.

This comparative analysis shows that our relations accurately describe important features of the harmonic spectrum and the evolution of its shape as the laser beam intensity increases. More specifically, for low intensities of the order of 10^{13} W/cm^2, the spectrum has the well-known plateau shape, followed by a sharp decrease at the cutoff edge. When the beam intensity is increased toward 10^{15} W/cm^2, the slope of the plateau increases so that the harmonic intensities decrease by 1 or 2 orders of magnitude. When the beam intensity is increased even more toward 10^{17} W/cm^2, there is a maximum at the left edge of the spectrum, which is followed by a strong decrease of many orders of magnitude, and the spectrum concentrates in a very limited domain, under a certain frequency, far lower than the theoretical cutoff frequency. These theoretical results are in good agreement with numerous other experimental data from the literature, for many atoms and ions, and for a large spectrum of laser beam intensities.

In Section B.4, we give the Mathematica 7 program, which has been used to calculate the harmonic spectra.

CONCLUSIONS

In Volume I we presented exact connections between quantum and classical equations for two different systems, namely, stationary atomic and molecular (AM) and electrodynamic (ED) systems composed of particles and ED fields. These systems have two common properties. First, basic mathematical equations lead, without any approximation, to wave functions associated to the classical solutions of the Hamilton–Jacobi and relativistic Hamilton–Jacobi equations, respectively, for AM and ED systems. Second, both types of systems have periodicity properties which lead to highly simplified models. The periodicity property for AM systems is related to the wave behavior of these systems, which results from the properties of the Schrödinger equation. On the other hand, in the case of the ED systems, we proved that all the expressions of the components of the electromagnetic fields resulted from interactions between very intense laser beams and electron plasmas or electron beams, as well as the expression of the high harmonic rate of generation in interactions with atoms, are periodic functions of only one variable, which is the phase of the incident electromagnetic field.

These properties lead to accurate models, in good agreement with the experimental data from literature. It is worth noting that the accuracy of the models developed for AM systems is comparable to the accuracy of the Hartree–Fock method. Moreover, the calculation models derived for ED systems lead to accurate angular and spectral distributions of the harmonics generated in the interactions between very intense electromagnetic fields and electron plasmas, relativistic electron beams, or atoms, in good agreement with numerous experimental data from literature.

Details of Calculation of the Correction Term E_{m1s}

We have the relations $\varepsilon_0\mu_0 = 1/c^2$, $|E_1| = (R_\infty Z_1^2)/n_1^2$, and $a_0 = \hbar^2/(mK_1)$, resulted from Eq. (B.10) from Volume I. The expressions of μ_1 and τ_1 are given, respectively, by Eqs. (1.9) and (B.11) from Volume I. Introducing these relations together with:

$$r_0 = \frac{e^2}{4\pi\varepsilon_0 mc^2} \quad \text{and} \quad \alpha = \sqrt{\frac{r_0}{a_0}} \tag{A.1}$$

in Eq. (1.20), we obtain:

$$
\begin{aligned}
E_{m1s} &= \frac{|E_1|}{4\pi n_1 \hbar} \cdot \frac{\mu_0}{4\pi}\left(\frac{e\hbar}{2m}\right)^2 \sqrt{\frac{mr_M}{2K_1 Z_1}} I = \frac{R_\infty Z_1^2}{16\pi n_1^3} \frac{\varepsilon_0 \mu_0 e^2}{4\pi\varepsilon_0 mm} \frac{\hbar}{\sqrt{\frac{mr_M}{2K_1 Z_1}}} I \\
&= \frac{R_\infty Z_1^{3/2}}{16\sqrt{2}\pi n_1^3} r_0 \frac{\hbar}{\sqrt{mK_1}}\sqrt{r_M} I = \frac{R_\infty Z_1^{3/2}}{16\sqrt{2}\pi n_1^3} r_0 \sqrt{a_0}\sqrt{r_M} I \\
&= \frac{R_\infty Z_1^{3/2}}{16\sqrt{2}\pi n_1^3} \frac{r_0^{3/2}\sqrt{r_M}}{\alpha} I
\end{aligned}
\tag{A.2}
$$

From Eq. (B.7) of Volume I, we have $\cos\theta = (1/e)[1 - (u/r)]$ which, together with Eq. (1.20), give the following expression for the integral I:

$$I = \int_{r_m}^{r_M}\left[\left(2 - \frac{3}{e^2}\right) + \frac{6u}{e^2}\frac{1}{r} - \frac{3u^2}{e^2}\frac{1}{r^2}\right]\frac{dr}{r^2\sqrt{(r_M - r)(r - r_m)}} \tag{A.3}$$

The above integral can be written in terms of the following three integrals:

$$I_1 = \int_{r_m}^{r_M}\frac{dr}{r^2\sqrt{(r_M - r)(r - r_m)}} = \frac{\pi}{2}\frac{r_m + r_M}{(r_m r_M)^{3/2}} \tag{A.4}$$

$$I_2 = \int_{r_m}^{r_M}\frac{dr}{r^3\sqrt{(r_M - r)(r - r_m)}} = \frac{\pi}{8}\cdot\frac{3r_m^2 + 2r_m r_M + 3r_M^2}{(r_m r_M)^{5/2}} \tag{A.5}$$

$$I_3 = \int_{r_m}^{r_M}\frac{dr}{r^4\sqrt{(r_M - r)(r - r_m)}} = \frac{\pi}{16}\frac{(r_m + r_M)(5r_m^2 - 2r_m r_M + 5r_M^2)}{(r_m r_M)^{7/2}} \tag{A.6}$$

Since the C_1 and C_2 curves are quasi-linear, we have $r_m \ll r_M$, and the expressions of I_1, I_2, and I_3 become:

$$I_1 \cong \frac{\pi}{2} \frac{1}{r_m^{3/2} \sqrt{r_M}}, \quad I_2 \cong \frac{3\pi}{8} \frac{1}{r_m^{5/2} \sqrt{r_M}}, \quad I_3 \cong \frac{5\pi}{16} \frac{1}{r_m^{7/2} \sqrt{r_M}} \quad (A.7)$$

Introducing the expressions of I_1, I_2, and I_3 in Eq. (A.3), we obtain:

$$I = \left(2 - \frac{3}{e^2}\right) \frac{\pi}{2} \frac{1}{r_m^{3/2} \sqrt{r_M}} + \frac{6u}{e^2} \frac{3\pi}{8} \frac{1}{r_m^{5/2} \sqrt{r_M}} - \frac{3u^2}{e^2} \frac{5\pi}{16} \frac{1}{r_m^{7/2} \sqrt{r_M}} \quad (A.8)$$

Since the C_1 and C_2 curves are quasi-linear, from Eq. (B.8) of Volume I, we have $u \cong 2r_m$ and $e \cong 1$, and I becomes:

$$I = \frac{\pi}{4r_m^{3/2} \sqrt{r_M}} \quad (A.9)$$

From Eqs. (A.2) and (A.9), we obtain:

$$E_{m1s} = \frac{R_\infty Z_1^{3/2}}{64\sqrt{2} n_1^3} \frac{1}{\alpha(r_m/r_0)^{3/2}} \quad (A.10)$$

We have to calculate the expression of r_m/r_0 from the relation between forces in the vicinity of nucleus. The expressions of the forces shown in Figure 1.2A are:

$$F_{r1} = \frac{ZK_1}{R^2} - \frac{mv_1^2}{r_c} \quad (A.11)$$

where the second term in the right hand member is the centrifugal force and r_c is the curvature radius of the electron trajectory.

$$F_{e1} = \frac{K_1}{(2u)^2} \quad (A.12)$$

$$F_{m1} = ev_1 B_2 \sin \gamma = ev_1 \frac{\mu_0}{4\pi} \frac{2\mu_2}{(2u)^3} \sin \gamma = K_2 \frac{v_1}{u^3} \quad (A.13)$$

It is easy to show that $\sin \gamma \cong 1/\sqrt{2}$ in the case of the elliptic quasi-pendular motion, when $r = u$ and $\theta = \pi/2$ (see Figure 1.2B). The expression of K_2 becomes:

$$K_2 = \frac{\mu_0}{4\pi} \frac{e^2 \hbar}{8\sqrt{2} m} \quad (A.14)$$

The following relations are valid (see Figure 1.2A):

$$F_{r1}^2 = F_{e1}^2 + F_{m1}^2, \quad \frac{F_{e1}}{F_{r1}} = \frac{u}{R}, \quad \text{and} \quad \frac{F_{e1}}{F_{m1}} = \frac{u}{\sqrt{R^2 - u^2}} \qquad \text{(A.15)}$$

The equation of the total energy is:

$$E = -\frac{2ZK_1}{R} + \frac{K_1}{2u} + mv_1^2 \qquad \text{(A.16)}$$

Since E is much smaller than any term of the left hand member, in the vicinity of the nucleus, we have:

$$mv_1^2 \cong \frac{2ZK_1}{R} - \frac{K_1}{2u} \qquad \text{(A.17)}$$

We observe that, in the configuration shown in Figure 1.2A, the Lorentz force has the maximum value in the vicinity of the nucleus. In order to verify the validity of assumption (a1.3), we have to check that even in this configuration, the attraction force of the nucleus is dominant. Indeed, we have $F_{r1}/F_{m1} = R/\sqrt{R^2 - u^2} > 1$ and the above property is proved.

Noting

$$r_c = kR \qquad \text{(A.18)}$$

we have from Eqs. (A.11), (A.12), (A.15), and (A.18):

$$mv_1^2 = k\left(\frac{ZK_1}{R} - \frac{K_1 R^2}{4u^3}\right) \qquad \text{(A.19)}$$

From Eqs. (A.17) and (A.19), we obtain:

$$Z\left(1 - \frac{2}{k}\right) + \frac{R}{2ku} = \frac{R^3}{4u^3} \qquad \text{(A.20)}$$

This equation leads to the following domain of variation for k: $2 \leq k < \infty$, which corresponds to $1 \leq R/u < (4Z)^{1/3}$. The left limit of these domains correspond to the case $R = u$, when no magnetic force acts, while the right limit corresponds to the case when the magnetic force is high, comparable, as order of magnitude, with the attraction force of the nucleus. We thus assume that R/u is very close to $(4Z)^{1/3}$, which gives:

$$R \cong u(4Z)^{1/3} \qquad \text{(A.21)}$$

From Eqs. (A.12), (A.13), and (A.15), we obtain:

$$v_1 = R\frac{K_1}{4K_2}\sqrt{1 - \frac{u^2}{R^2}} \tag{A.22}$$

From Eqs. (A.17) and (A.22), we have:

$$v_1 = \frac{1}{\sqrt{R}}\sqrt{\frac{K_1}{m}}\sqrt{2Z - \frac{1}{2}\frac{R}{u}} = R\frac{K_1}{4K_2}\sqrt{1 - \frac{u^2}{R^2}} \tag{A.23}$$

or, taking into account Eqs. (A.21) and (A.23):

$$R^{3/2} = \frac{4K_2}{\sqrt{mK_1}}\sqrt{\frac{2Z - \frac{1}{2}(4Z)^{1/3}}{1 - (4Z)^{-2/3}}} = \frac{4K_2\sqrt{2Z}}{\sqrt{mK_1}} \tag{A.24}$$

Since the curve C_1 is quasi-linear, we have $u = 2r_m$ and from Eq. (A.21), we obtain:

$$R = u(4Z)^{1/3} = 2r_m(4Z)^{1/3} \tag{A.25}$$

From Eqs. (A.24) and (A.25), we have:

$$r_m^{3/2} = \frac{K_2}{\sqrt{mK_1}} \tag{A.26}$$

Taking into account the expressions of K_1, a_0, r_0, α, and K_2 from Eqs. (B.3), (B.10), of Volume I and (A.1), (A.14), together with the relation $\varepsilon_0\mu_0 = 1/c^2$, we have:

$$\frac{K_2}{\sqrt{mK_1}} = \frac{\mu_0}{4\pi}\frac{e^2\hbar}{8\sqrt{2}m}\frac{1}{\sqrt{mK_1}} = \frac{1}{8\sqrt{2}}\frac{\varepsilon_0\mu_0 e^2}{4\pi\varepsilon_0 m}\frac{\hbar}{\sqrt{mK_1}} = \frac{1}{8\sqrt{2}\alpha}r_0^{3/2} \tag{A.27}$$

From Eqs. (A.26) and (A.27), we obtain:

$$\left(\frac{r_m}{r_0}\right)^{3/2} = \frac{1}{8\sqrt{2}\alpha} \tag{A.28}$$

From Eqs. (A.10) and (A.28), we obtain the final relation of the interaction energy between the magnetic dipole moments of the electrons:

$$E_{m1s} = \frac{R_\infty Z_1^{3/2}}{8n_1^3} \tag{A.29}$$

Mathematica 7 Programs

B.1 PROGRAMS FOR THE CALCULATION OF THE ATOMIC PROPERTIES

We have the following equivalences between the notations from the book, and the symbols used in programs: $Z \equiv Z$, $s_{31ei} \equiv s31ei$, $s_{31ef} \equiv s31ef$, $s_{13e} \equiv s13e$, $s_{51ei} \equiv s51ei$, $s_{51ef} \equiv s51ef$, $s_{15e} \equiv s13e$, $\underline{r}_{M1} \equiv rM1$, $\underline{r}_{M3} \equiv rM3$, $\underline{r}_{m3} \equiv rm3$, $\underline{r}_{M5} \equiv rM5$, $\varrho \equiv r$, $\underline{E}_1 \equiv E1$, $\underline{E}_3 \equiv E3$, $\underline{E}_5 \equiv E5$, $\underline{E}_{m1s} \equiv Em1s$, $\underline{E}_{m2s} \equiv Em2s$, $E \equiv Et$, $P \equiv number\ \pi$.

B.1.1 The Lithium Atom in State $1s^2 2s$

Output data: s_{31ef}, s_{13e}, E

```
Z = 3;
P = 3.14159265359;
s31ei = .854942;
rM1 = 1/(Z-1/4);
rM3 = 4/(Z-2 s31ei);
c = 1/(P (1 + (rM1/rM3)^2));
f = c r^(1/2)/(((rM1/2)^2 + r^2)(rM3-r))^(1/2);
s31ef = NIntegrate[f,{r,0,rM3}]
s13e = s31ef (rM1/rM3)^3
E1 = -(Z-1/4-s13e)^2;
E3 = -(1/4)(Z-2 s31ef)^2;
Em1s = (Z-1/4-s13e)^(3/2)/8;
Et = 2 E1 + E3 + Em1s
0.854942
0.00137926
-14.9563
```

B.1.2 The Lithium Atom in State $1s^2 2p$

Output data: s_{31ef}, s_{13e}, E

```
Z = 3;
P = 3.14159265359;
s31ei = .979092;
rM1 = 1/(Z-1/4);
```

```
rM3 = (2/(Z-2 s31ei))(1 + 0.5^(1/2));
rm3 = (2/(Z-2 s31ei))(1-0.5^(1/2));
c = 1/(P (1 + (rM1/(rM3 + rm3))^2));
f = c r/(((rM1/2)^2 + r^2)(rM3-r)(r-rm3))^(1/2);
s31ef = NIntegrate[f,{r,rm3,rM3}]
s13e =  s31ef (rM1/(rm3 + rM3))^3
E1 = -(Z-1/4-s13e)^2;
E3 = -(1/4)(Z-2 s31ef)^2;
Em1s = (Z-1/4-s13e)^(3/2)/8;
Et = 2 E1 + E3 + Em1s
0.979092
0.000831799
-14.8174
```

B.1.3 The Beryllium Atom

Output data: s_{31ef}, s_{13e}, E

```
Z = 4;
P = 3.14159265359;
s31ei = .83882;
rM1 = 1/(Z-1/4);
rM3 = 4/(Z-1/4-2 s31ei);
c = 1/(P (1 + (rM1/rM3)^2));
f = c r^(1/2)/(((rM1/2)^2 + r^2)(rM3-r))^(1/2);
s31ef = NIntegrate[f,{r,0,rM3}]
s13e =  s31ef (rM1/rM3)^3
E1 = -(Z-1/4-2 s13e)^2;
E3 = -(1/4)(Z-2 s31ef -1/4)^2;
Em1s = (Z-1/4-2 s13e)^(3/2)/8;
Em2s = (Z-1/4-2 s31ef)^(3/2)/64;
Et = 2 E1 + Em1s + 2 E3 + Em2s
0.83882
0.00221203
-29.2533
```

B.1.4 The Boron Atom
Output data: s_{31ef}, s_{13e}, E

```
Z = 5;
P = 3.14159265359;
s31ei = .862983;
rM1 = 1/(Z-1/4);
rM3 = 2(1 + (1-1/18)^(1/2))/(Z-2 s31ei-1/3^(1/2));
rm3 = 2(1-(1-1/18)^(1/2))/(Z-2 s31ei-1/3^(1/2));
c = 1/(P (1 + (rM1/(rM3 + rm3))^2));
f = c r/((r^2 + (rM1/2)^2)(rM3-r)(r-rm3))^(1/2);
s31ef = NIntegrate[f,{r,rm3,rM3}]
s13e = s31ef (rM1/(rM3 + rm3))^3
E1 = -(Z-1/4-3 s13e)^2;
E3 = -(1/4) (Z-2 s31ef-1/3^(1/2))^2;
Em1s =  (Z-1/4-3 s13e)^(3/2)/8;
Et = 2 E1 + 3 E3 + Em1s
0.862983
0.00246735
-49.1475
```

B.1.5 The Carbon Atom
Output data: s_{31ef}, s_{13e}, E

```
Z = 6;
P = 3.14159265359;
s31ei = .85505;
rM1 = 1/(Z-1/4);
rM3 = 2(1 + (1-3/64)^(1/2))/(Z-2 s31ei-(3 3^(1/2))/(4 2^(1/2)));
rm3 = 2(1-(1-3/64)^(1/2))/(Z-2 s31ei-(3 3^(1/2))/(4 2^(1/2)));
c = 1/(P (1 + (rM1/(rM3 + rm3))^2));
f = c r/((r^2 + (rM1/2)^2)(rM3-r)(r-rm3))^(1/2);
s31ef = NIntegrate[f,{r,rm3,rM3}]
s13e = s31ef (rM1/(rM3 + rm3))^3
E1 = -(Z-1/4-4 s13e)^2;
E3 = -(1/4)(Z-2 s31ef-(3 3^(1/2))/(4 2^(1/2)))^2;
Em1s = (Z-1/4-4 s13e)^(3/2)/8;
Et = 2 E1 + 4 E3 + Em1s
0.85505
0.00269287
-75.5248
```

B.1.6 The Oxygen atom

Output data: s_{31ef}, s_{13e}, E

Z = 8;
P = 3.14159265359;
s31ei = .839822;
rM3 = 2(1 + (1-3/144)^(1/2))/(Z-2 s31ei-2^(1/2)-1/4);
rm3 = 2(1-(1-3/144)^(1/2))/(Z-2 s31ei-2^(1/2)-1/4);
rM1 = 1/(Z-1/4);
c = 1/(P (1 + (rM1/(rM3 + rm3))^2));
f = c r/((rM3-r)(r-rm3)(r^2 + (rM1/2)^2))^(1/2);
s31ef = NIntegrate[f,{r,rm3,rM3}]
s13e = s31ef (rM1/(rM3 + rm3))^3
E1 = -(Z-1/4-6 s13e)^2;
E3 = -(1/4)(Z-2 s31ef-2^(1/2)-1/4)^2;
Em1s = (Z-1/4-6 s13e)^(3/2)/8;
Et = 2 E1 + 6 E3 + Em1s
0.839822
0.00284565
-149.428

B.1.7 The Nitrogen Atom

Output data: s_{31ef}, s_{13e}, s_{51ef}, s_{15e}, E

Z = 7;
P = 3.14159265359;
s31ei = .830382;
rM1 = 1/(Z-1/4);
rM3 = 4/(Z-2 s31ei-1/4-3/(2 2^(1/2)));
c = 1/(P (1 + (rM1/rM3)^2));
f = c r^(1/2)/((rM3-r)(r^2 + (rM1/2)^ 2))^(1/2);
s31ef = NIntegrate[f,{r,0,rM3}]
s13e = s31ef (rM1/rM3)^3
s51ei = .829599;
rM5 = 4/(Z-2 s51ei-1/2^(1/2)-1/3^(1/2));
c = 1/(P (1 + (rM1/rM5)^2));
f = c r^(1/2)/((rM5-r)(r^2 + (rM1/2)^2))^(1/2);
s51ef = NIntegrate[f,{r,0,rM5}]
s15e = s51ef (rM1/rM5)^3
E1 = -(Z-1/4-2 s13e-3 s15e)^2;
E3 = -(1/4) (Z-2 s31ef-1/4-3/(2 2^(1/2)))^2;
E5 = -(1/4) (Z-2 s51ef-1/2^(1/2)-1/3^(1/2))^2;

Em1s = (Z-1/4-2 s13e-3 s15e)^(3/2)/8;
Et = 2 E1 + 2 E3 + 3 E5 + Em1s
0.830382
0.0027583
0.829599
0.00281308
-109.018

B.2 PROGRAMS FOR THE CALCULATION OF THE MOLECULAR PROPERTIES

We have the following equivalences between the notations from the book and the symbols used in programs: $Z_A \equiv ZA$, $Z_B \equiv ZB$, $s_{1,A1} \equiv s1A1$, $s_{A1,1} \equiv sA11$, $s_{A1,A2} \equiv sA1A2$, $s_{1,2} \equiv s12$, $s_{1,A3} \equiv s1A3$, $s_{1,A4} \equiv s1A4$, $s_{1,A5} \equiv s1A5$, $s_{A3,A4} \equiv sA3A4$, $s_{1,2(A)} \equiv s12A$, $s_{1,2(B)} \equiv s12B$, $\alpha \equiv alpha$, $\pi\alpha/180 \equiv al$, $Z_1 \equiv Z1$, $Z_{1M} \equiv Z1M$, $Z_{A1} \equiv ZA1$, $Z_{A3} \equiv ZA3$, $Z_{A4} \equiv ZA4$, $Z_{A5} \equiv ZA5$, $Z_{1(A)} \equiv Z1A$, $Z_{1M(A)} \equiv Z1MA$, $Z_{1(B)} \equiv Z1B$, $Z_{1M(B)} \equiv Z1MB$, $Z_{2(B)} \equiv Z2B$, $Z_{2M(B)} \equiv Z2MB$, $Z_{nA} \equiv ZnA$, $Z_{nB} \equiv ZnB$, $e \equiv excen$, $\underline{r}_{A3} \equiv rA3$, $\underline{r} \equiv r$, $\underline{b} \equiv b$, $\underline{d} \equiv d$, $a_{cA} \equiv acA$, $a_{cB} \equiv acB$, $t_{cA} \equiv tcA$, $t_{cB} \equiv tcB$, $\underline{E} \equiv Et$, $\underline{E}_b \equiv Eb$, $\underline{E}_{b1} \equiv Eb1$, $\underline{E}_{b2} \equiv Eb2$, $r_0 \equiv r0$, $\mu_e \equiv me$, $\underline{\sigma} \equiv s$, $\underline{a} \equiv a$, $\underline{\sigma}_1 \equiv s1$, $\underline{\sigma}_2 \equiv s2$, $\underline{T}_m \equiv T$, $P \equiv number \ \pi$.

B.2.1 The Li_2 Molecule
Output data: E, r_0

Input data.

ZA = 3;
s1A1 = 0.854942;
sA11 = 0.0013792;
sA1A2 = .25;
Z1 = ZA-2*s1A1;
Z1M = ZA-2;
ZnA = ZA-2;
Et = Eb-4*(ZA-sA1A2-sA11)^2 + (ZA-sA1A2-sA11)^(3/2)/4;
r0 = 2*s*2*.529177 A;
System of equations
Eq1 = Eb + Z1^2/2 + (8*Z1M)/(9*a^2 + 25*s^2)^(1/2)-2
/(9*a^2 + s^2)^(1/2)
-ZnA^2/(2*s);
Eq2 = Eb + (4 (Z1M)^(1/2))/(s^2 + a^2)^(3/4);

Eq3 = Eb + (4*Z1M)/(s^2 + a^2)^(1/2)-ZnA^2/(2*s)-1/(2 a);
The solution of the system of equations
Sol = FindRoot[{Eq1 == 0,Eq2 == 0,Eq3 == 0},{{Eb,1},{s,1.2},
{a,2.5}}]
{Eb- > -1.0115,s- > 1.46474,a- > 2.02683}
Verification.
N[Eq1]/.Sol
N[Eq2]/.Sol
N[Eq3]/.Sol
-1.11022*10^-16
0.
0.
Calculation of other quantities.
Et/.Sol
r0/.Sol
-30.0919
3.10043 A

B.2.2 The Be_2 Molecule
Output data: E, r_0

Input data.

ZA = 4;
s1A1 = 0.83882;
sA11 = .00221203;
s1A3 = 2^(1/2)/(4*(1 + s/(s^2 + a^2)^(1/2))^(1/2));
sA1A2 = .25;
Z1 = ZA-2 s1A1-s1A3;
Z1M = ZA-2-s1A3;
ZA3 = Z1;
ZA1 = ZA-sA1A2-2*sA11;
rA3 = 3/ZA3;
ZnA = ZA-2-(2 s)^2/(2 s + rA3)^2;
Et = -(Z1^2/2)-ZA3^2/2-4*ZA1^2 + 1/(2((1/4*s)^2 + (3/4*a)^2)^(1/2))
-(2*(ZA-2))/((5/4*s)^2 + (3/4*a)^2)^(1/2) + 2/((5/4*s + rA3)^2 + (3/
4*a)^2)^(1/2)
-(2*(ZA-2))/(2*s + rA3) + 1/(2*s + 2*rA3) + (ZA-2)^2/(2*s) + ZA1^
(3/2)/4;
r0 = 2*s*2*.529177 A;

System of equations.

Eq1 = Eb + Z1^2/2 + (8*Z1M)/(9*a^2 + 25*s^2)^(1/2)-2
/(9*a^2 + s^2)^(1/2)
-ZnA^2/(2*s);
Eq2 = Eb + (4 Z1M^(1/2))/(s^2 + a^2)^(3/4);
Eq3 = Eb + (4*Z1M)/(s^2 + a^2)^(1/2)-ZnA^2/(2*s)-1/(2 a);

The solution of the system of equations.

Sol = FindRoot[{Eq1 == 0,Eq2 == 0,Eq3 == 0},{{Eb,-2},{s,1},
{a,1}}]
{Eb- > -2.39104,s- > 1.04502,a- > 1.32709}

Verification.

N[Eq1]/.Sol
N[Eq2]/.Sol
N[Eq3]/.Sol
-4.44089*10^-16
0.
-8.88178*10^-16

Calculation of other quantities.

Et/.Sol
r0/.Sol
-58.7676
2.212 A

B.2.3 The B_2 Molecule

Output data: E, r_0

Input data.
ZA = 5;
P = 3.14159265359;
s1A1 = 0.862983;
sA11 = .00246735;
sA1A2 = .25;
alpha = 63;
al = alpha*P/180;
s1A3 = 1/(4*(1/2 + (s*Cos[al])/(2*(s^2 + a^2)^(1/2)))^(1/2));
sA3A4 = 1/(4*Sin[al]);
Z1 = ZA-2*s1A1-2*s1A3;
Z1M = ZA-2-2*s1A3;
ZA1 = ZA-sA1A2-3*sA11;
ZA3 = ZA-2*s1A1-s1A3-sA3A4;

excen = 0.97;

r = 2/ZA3*(1 + excen^2/2);

d = r*Cos[al];

b = r*Sin[al];

ZnA = ZA-2-(2*(2 s)^2*(2*s + d))/((2* s + d)^2 + b^2)^(3/2);

Et = -(Z1^2/2)-ZA3^2-4*ZA1^2 + 1/(2*((1/4*s)^2 + (3/4*a)^2)^(1/2))

-(2*(ZA-2))/((5/4*s)^2 + (3/4*a)^2)^(1/2) + 4/((5/4*s + d)^2 + (3/4*a)

^2 + b^2)^(1/2)

-(4*(ZA-2))/((d + 2*s)^2 + b^2)^(1/2) + 1/((s + d)^2 + b^2)^(1/2) +

1/(s + d)

 + (ZA-2)^2/(2*s) + ZA1^(3/2)/4;

r0 = 2*s*2*.529177 A;

System of equations.

Eq1 = Eb + Z1^2/2 + (8*Z1M)/(9*a^2 + 25*s^2)^(1/2)-2

/(9*a^2 + s^2)^(1/2)

-ZnA^2/(2*s);

Eq2 = Eb + (4 Z1M^(1/2))/(s^2 + a^2)^(3/4);

Eq3 = Eb + (4*Z1M)/(s^2 + a^2)^(1/2)-ZnA^2/(2*s)-1/(2 a);

The solution of the system of equations.

Sol = FindRoot[{Eq1 == 0,Eq2 == 0,Eq3 == 0},{{Eb,1},{s,1},

{a,1}}]

{Eb- > -3.87125,s- > 0.837345,a- > 1.07618}

Verification.

N[Eq1]/.Sol

N[Eq2]/.Sol

N[Eq3]/.Sol

-1.77636*10^-15

8.88178*10^-16

-1.77636*10^-15

Calculation of other quantities.

Et/.Sol

r0/.Sol

-98.3291

1.77241 A

B.2.4 The C_2 Molecule with Single Bond

Output data: E, r_0

Input data.

ZA = 6;

P = 3.14159265359;
s1A1 = 0.85505;
sA11 = .00269287;
alpha = 65;
al = alpha*P/180;
s1A3 = 2^(1/2)/(4*(1 + (s*Cos[al])/(s^2 + a^2)^(1/2))^(1/2));
s1A4 = 2^(1/2)/(4*(1 + (s*Cos[al]-3^(1/2)/2*a*Sin[al])/(s^2 + a^2)^(1/2))^(1/2));
s1A5 = 2^(1/2)/(4*(1 + (s*Cos[al] + 3^(1/2)/2*a*Sin[al])/(s^2 + a^2)^(1/2))^(1/2));
sA3A4 = 1/(2*3^(1/2)*Sin[al]);
sA1A2 = .25;
Z1 = ZA-2*s1A1-s1A3-s1A4-s1A5;
Z1M = ZA-2-s1A3-s1A4-s1A5;
ZA1 = ZA-sA1A2-4*sA11;
ZA3 = ZA-2*s1A1-s1A3-2*sA3A4;
ZA4 = ZA-2*s1A1-s1A4-2*sA3A4;
ZA5 = ZA-2*s1A1-s1A5-2*sA3A4;
r = 1/ZA3 + 1/ZA4 + 1/ZA5;
d = r*Cos[al];
b = r*Sin[al];
ZnA = ZA-2-(3*(2*s)^2*(2*s + d))/((2* s + d)^2 + b^2)^(3/2);
Et = -(Z1^2/2)-ZA3^2/2-ZA4^2/2-ZA5^2/2-4*ZA1^2
 + 1/(2*((1/4*s)^2 + (3/4*a)^2)^(1/2))-(2*(ZA-2))/((5/4*s)^2 + (3/4*a)^2)^(1/2)
 + 2/((5/4*s + d)^2 + (3/4*a)^2 + b^2)^(1/2)
 + 2/((5/4*s + d)^2 + (3/4*a + 3^(1/2)/2*b)^2 + (b/2)^2)^(1/2)
 + 2/((5/4*s + d)^2 + (3/4*a-3^(1/2)/2*b)^2 + (b/2)^2)^(1/2)
-(6*(ZA-2))/((2*s + d)^2 + b^2)^(1/2) + 3/((2*s + 2*d)^2 + (2*b)^2)^(1/2)
 + 6/((2*s + 2*d)^2 + b^2)^(1/2) + (ZA-2)^2/(2*s) + 2*ZA1^(3/2)/8;
r0 = 2*s*2*.52918 A;
System of equations.
Eq1 = Eb + Z1^2/2 + (8*Z1M)/(9*a^2 + 25*s^2)^(1/2)
-2/(9*a^2 + s^2)^(1/2)-ZnA^2/(2*s);
Eq2 = Eb + (4(Z1M)^(1/2))/(s^2 + a^2)^(3/4);
Eq3 = Eb + (4*Z1M)/(s^2 + a^2)^(1/2)-ZnA^2/(2*s)-1/(2 a);
The solution of the system of equations.
Sol = FindRoot[{Eq1 == 0,Eq2 == 0,Eq3 == 0},{{Eb,-5.6},
{s,.75},{a,.85}}]

{Eb- > -5.69164,s- > 0.751248,a- > 0.857921}
Verification.
N[Eq1]/.Sol
N[Eq2]/.Sol
N[Eq3]/.Sol
-8.88178*10^-16
1.77636*10^-15
2.66454*10^-15
Calculation of other quantities.
Et/.Sol
r0/.Sol
-150.365
1.59018 A

B.2.5 The C_2 Molecule with Double Bond

Output data: E, r_0

Input data.
ZA = 6;
P = 3.14159265359;
s1A1 = 0.85505;
sA11 = .00269287;
sA1A2 = .25;
alpha = 41;
al = alpha*P/180;
s12 = (s^2 + a^2)^(1/2)/(4*a);
s1A3 = 2^(1/2)/(4*(1 + (s*Cos[al]-2^(1/2)/2*a*Sin[al])/(s^2 + a^2)^
(1/2))^(1/2));
s1A4 = 2^(1/2)/(4*(1 + (s*Cos[al] + 2^(1/2)/2*a*Sin[al])/(s^2 + a^2)^
(1/2))^(1/2));
sA3A4 = 1/(4*Sin[al]);
Z1 = ZA-2*s1A1-s12-s1A3-s1A4;
Z1M = ZA-2-s12-s1A3-s1A4;
ZA1 = ZA-sA1A2-4*sA11;
ZA3 = ZA-2*s1A1-sA3A4-s1A3-s1A4;
ZA4 = ZA3;
excen = .98;
r = 2/ZA3*(1 + excen^2/2);
d = r*Cos[al];
b = r*Sin[al];

ZnA = ZA-2-(2*(2*s)^2*(2*s + d))/((2* s + d)^2 + b^2)^(3/2);
Et = -Z1^2-ZA3^2-4*ZA1^2 + 4/((s/2)^2 + 2*((3*a)/4)^2)^(1/2)
-(4*(ZA-2))/((5/4*s)^2 + (3/4*a)^2)^(1/2)
  + 4/(((5*s)/4 + d)^2 + (2^(1/2)/2*b)^2 + ((3*a)/4-2^(1/2)/2*b)^2)^(1/2)
  + 4/(((5*s)/4 + d)^2 + (2^(1/2)/2*b)^2 + ((3*a)/4 + 2^(1/2)/2*b)^2)^
(1/2)
-(4*(ZA-2))/((2*s + d)^2 + b^2)^(1/2) + 1/((s + d)^2 + b^2)^(1/2) +
1/(s + d)
  + (ZA-2)^2/(2*s) + ZA1^(3/2)/4;
r0 = 2*s*2*.52918A;
System of equations.
Eq1 = Eb + Z1^2 + (16*(Z1M + s12))/(25 s^2 + 9 a^2)^(1/2)-8/(s^2 +
4.5*a^2)^(1/2)
-ZnA^2/(2*s);
Eq2 = Eb + (8*(Z1M)^(1/2))/(s^2 + a^2)^(3/4);
Eq3 = Eb + (8*(Z1M + s12))/(s^2 + a^2)^(1/2)-1/a-(2*2^(1/2))/a-
ZnA^2/(2*s);
The solution of the system of equations.
Sol = FindRoot[{Eq1 == 0,Eq2 == 0,Eq3 == 0},{{Eb,-12},{s,.6},
{a,.9}}]
{Eb- > -12.1816,s- > 0.641059,a- > 0.894043}
Verification.
N[Eq1]/.Sol
N[Eq2]/.Sol
N[Eq3]/.Sol
-3.55271*10^-15
0.
3.55271*10^-15
Calculation of other quantities.
Et/.Sol
r0/.Sol
-151.148
1.35694 A

B.2.6 The C_2 Molecule with Triple Bond
Output data: E, r_0

Input data.
ZA = 6;
s1A1 = 0.85505;

sA11 = .00269287;

sA1A2 = .25;

s12 = 2^(1/2)/4/(1-(s^2-1/2*a^2)/(s^2 + a^2))^(1/2);

s1A3 = 2^(1/2)/4/(1 + s/(s^2 + a^2)^(1/2))^(1/2);

Z1 = ZA-2*s1A1-2*s12-s1A3;

Z1M = ZA-2-2*s12-s1A3;

ZA1 = ZA-sA1A2-4*sA11;

ZA3 = ZA-2*s1A1-3*s1A3;

excen = .98;

rA3 = 2/ZA3*(1 + excen^2/2);

ZnA = ZA-2-(2*s)^2/(2* s + rA3)^2;

Et = -6*Z1^2/4-2*ZA3^2/4-4*ZA1^2 + 3/((1/2*s)^2 + (3/2*a)^2)^(1/2)
+ 6/((1/2*s)^2 + (3/4*a)^2)^(1/2)-(6*(ZA-2))/((5/4*s)^2 + (3/4*a)^2)^
(1/2)
+ 6/(((5*s)/4 + rA3)^2 + ((3*a)/4)^2)^(1/2)-(2*(ZA-2))/(2*s + rA3)
+ 1/(2*s + 2*rA3) + (ZA-2)^2/(2*s) + ZA1^(3/2)/4;

r0 = 2*s*2*.52918A;

System of equations.

Eq1 = Eb + 6*Z1^2/4 + (24*(Z1M + 2*s12))/(9*a^2 + 25*s^2)^(1/2)-
6/(s^2 + 9*a^2)^(1/2)-24/(4*s^2 + 9*a^2)^(1/2)-ZnA^2/(2*s);

Eq2 = Eb + (12*(Z1M)^(1/2))/(s^2 + a^2)^(3/4);

Eq3 = Eb + (12*(Z1M + 2*s12))/(s^2 + a^2)^(1/2)-6/(a*3^(1/2))-6/a-
3/(2*a)
-ZnA^2/(2*s);

The solution of the system of equations.

Sol = FindRoot[{Eq1 == 0,Eq2 == 0,Eq3 == 0},{{Eb,12},{s,.6},
{a,.9}}]

{Eb- > -17.1499,s- > 0.639379,a- > 0.942817}

Verification.

N[Eq1]/.Sol

N[Eq2]/.Sol

N[Eq3]/.Sol

3.55271*10^-15

-3.55271*10^-15

-3.55271*10^-15

Calculation of other quantities.

Et/.Sol

r0/.Sol

-150.923

1.35339 A

B.2.7 The LiH Molecule
Output data: E, r_0, μ_e

Input data.
ZA = 3;
ZB = 1;
P = 3.14159265359;
s1A1 = 0.854942;
sA11 = .0013792;
sA1A2 = .25;
s12A = (s1^2 + a^2)^(1/2)/(4 a);
s12B = (s2^2 + a^2)^(1/2)/(4 a);
Z1A = ZA-2*s1A1-s12A;
Z1MA = ZA-2-s12A;
ZA1 = ZA-sA1A2-2*sA11;
Z1B = ZB-s12B;
Z1MB = ZB-s12B;
ZnA = 1;
ZnB = 1;
acA = 1-(T*(s1^2 + a^2)^(1/2))/Z1MA;
acB = 1-(T*(s2^2 + a^2)^(1/2))/Z1MB;
IA = ArcSin[(acA)^(1/2)]-(acA)^(1/2)(1-acA)^(1/2);
IB = ArcSin[(acB)^(1/2)]-(acB)^(1/2)(1-acB)^(1/2);
tcA = (2 IA)/(P acA^(3/2));
tcB = (2 IB)/(P acB^(3/2));
Et = Eb-2*ZA1^2 + ZA1^(3/2)/8;
r0 = (s1 + s2)*.529177*2 A;
me = -5.08315*(-2*s1 + ZB*(s1 + s2)) D;
System of equations.
Eq1 = Eb + Z1A^2/2 + (8(Z1MB + s12B))/(9 a^2 + (s1 + 4 s2)^2)^
(1/2)
-(ZnA ZnB)/(s1 + s2);
Eq2 = Eb + 2 Z1B^2 + (8(Z1MA + s12A))/(9 a^2 + (s2 + 4 s1)^2)^
(1/2)
-(ZnA ZnB)/(s1 + s2);
Eq3 = Eb + (4(Z1MA)^(1/2))/(tcA (s1^2 + a^2)^(3/4));
Eq4 = Eb + (2(Z1MB)^(1/2))/(tcB (s2^2 + a^2)^(3/4));
Eq5 = Eb + (2(Z1MA + s12A))/(s1^2 + a^2)^(1/2)
 + (2(Z1MB + s12B))/(s2^2 + a^2)^(1/2)-1/(2 a)-(ZnA ZnB)/
(s1 + s2)-2 T;

The solution of the system of equations.
Sol = FindRoot[{Eq1 == 0,Eq2 == 0,Eq3 == 0,Eq4 == 0,
Eq5 == 0},
{{Eb,1.6},{s1,1.5},{s2,.25},{a,1.2},{T,.04}}]
{Eb- > -1.65659,s1- > 1.47713,s2- > 0.267425,a- > 1.18038,T-
> 0.0284159}
Verification.
N[Eq1]/.Sol
N[Eq2]/.Sol
N[Eq3]/.Sol
N[Eq4]/.Sol
N[Eq5]/.Sol
4.44089*10^-16
0.
-2.22045*10^-16
-8.88178*10^-16
6.66134*10^-16
Calculation of other quantities.
Et/.Sol
r0/.Sol
me/.Sol
-16.1821
1.84635 A
6.1491 D

B.2.8 The BeH Molecule
Output data: E, r_0, μ_e

Input data.
ZA = 4;
ZB = 1;
P = 3.14159265359;
T = .015;
s1A1 = 0.83882;
sA11 = .00221203;
sA1A2 = .25;
s12A = (s1^2 + a^2)^(1/2)/(4 a);
s12B = (s2^2 + a^2)^(1/2)/(4 a);
s1A3 = 2^(1/2)/(4*(1 + s1/(s1^2 + a^2)^(1/2))^(1/2));
Z1A = ZA-2* s1A1-s12A-s1A3;

Z1MA = ZA-2-s12A-s1A3;
ZA1 = ZA-sA1A2-3*sA11;
ZA3 = ZA-2*s1A1-2*s1A3;
Z1B = ZB-s12B;
Z1MB = ZB-s12B;
rA3 = 3/(ZA-2* s1A1);
ZnA = ZA-2-(s1 + s2)^2/(s1 + s2 + rA3)^2;
ZnB = 1;
acA = 1-(T (s1^2 + a^2)^(1/2))/Z1MA;
acB = 1-(T (s2^2 + a^2)^(1/2))/Z1MB;
IA = ArcSin[(acA)^(1/2)]-(acA)^(1/2)(1-acA)^(1/2);
IB = ArcSin[(acB)^(1/2)]-(acB)^(1/2)(1-acB)^(1/2);
tcA = (2 IA)/(P acA^(3/2));
tcB = (2 IB)/(P acB^(3/2));
Et = -(ZA3^2/4)-2 *ZA1^2-(2*(Z1MA + s12A))/(s1^2 + a^2)^(1/2)
-(2*(Z1MB + s12B))/(s2^2 + a^2)^(1/2) + 1/(2*a)
-1/(rA3 + s1 + s2) + ((ZA-2)*ZB)/(s1 + s2) + ZA1^(3/2)/8 + 2*T;
r0 = (s1 + s2)*.529177*2 A;
me = -5.08315*(-2*s1 + ZB*(s1 + s2) + rA3) D;
System of equations.
Eq1 = Eb1 + Z1A^2/2 + (8 (Z1MB + s12B))/(9 a^2 + (s1 + 4 s2)^2)^
(1/2)
-(ZnA ZnB)/(s1 + s2);
Eq2 = Eb2 + 2 Z1B^2 + (8 (Z1MA + s12A))/(9 a^2 + (s2 + 4 s1)^2)^
(1/2)
-(ZnA ZnB)/(s1 + s2);
Eq3 = Eb1 + (4(Z1MA)^(1/2))/(tcA (s1^2 + a^2)^(3/4));
Eq4 = Eb2 + (2(Z1MB)^(1/2))/(tcB (s2^2 + a^2)^(3/4));
Eq5 = 0.5*(Eb1 + Eb2) + (2 (Z1MA + s12A))/(s1^2 + a^2)^(1/2)
 + (2   (Z1MB + s12B))/(s2^2 + a^2)^(1/2)-1/(2   a)-(ZnA   ZnB)/
(s1 + s2)-2 T;
The solution of the system of equations.
Sol = FindRoot[{Eq1 == 0,Eq2 == 0,Eq3 == 0,Eq4 == 0,
Eq5 == 0},
{{Eb1,-2.3},{Eb2,-2},{s1,1.35},{s2,.01},{a,1}}]
{Eb1- > -2.51578,Eb2- > -2.11007,s1- > 1.34462,s2- > 0.0149424,a-
> 0.979054}
Verification.
N[Eq1]/.Sol
N[Eq2]/.Sol

N[Eq3]/.Sol
N[Eq4]/.Sol
N[Eq5]/.Sol
-8.88178*10^-16
0.
1.33227*10^-15
4.44089*10^-16
0.
Output data.
Et/.Sol
r0/.Sol
me/.Sol
-30.4238
1.4389 A
0.192613 D

B.2.9 The BH Molecule
Output data: E, r_0, μ_e

Input data.
ZA = 5;
ZB = 1;
P = 3.14159265359;
s1A1 = 0.862983;
sA11 = .00246735;
sA1A2 = .25;
alpha = 63;
al = alpha*P/180;
s1A3 = 1/(4*(1/2 + (s1*Cos[al])/(2*(s1^2 + a^2)^(1/2)))^(1/2));
sA3A4 = 1/(4*Sin[al]);
Z1A = ZA-2*s1A1-2*s1A3;
Z1MA = ZA-2-2*s1A3;
Z1MB = 1;
ZA1 = ZA-sA1A2-3*sA11;
ZA3 = ZA-2*s1A1-s1A3-sA3A4;
ZA4 = ZA3;
Z2B = ZB;
Z2MB = ZB;
excen = .97;
r = 2/ZA3*(1 + excen^2/2);

d = r*Cos[al];
b = r*Sin[al];
ZnA = ZA-2-(2*(s1 + s2)^2*(s1 + s2 + d))/((s1 + s2 + d)^2 + b^2)^
(3/2);
ZnB = 1;
Et = -(Z1A^2/4)-1-ZA3^2/2-2*ZA1^2 + 1/((s1/4 + s2/4)^2 + (3/2*a)
^2)^(1/2)
-1 /((s1/4 + s2)^2 + (3/4 a)^2)^(1/2)-(ZA-2)/((s2/4 + s1)^2 + (3/4 a)^2)
^(1/2)
 + 2/((s2/4 + s1 + d)^2 + (3/4 a)^2 + b^2)^(1/2)-2/((s1 + s2 + d)^2 + b^2)
^(1/2)
 + (ZA-2)/(s1 + s2) + ZA1^(3/2)/8;
r0 = (s1 + s2)*.529177*2A;
me = -5.08315*(-2*s1 + ZB*(s1 + s2) + 2*d) D;
System of equations.
Eq1 = Eb + Z1A^2/4 + Z2B^2 + (4*Z1MB)/(9*a^2 + (s1 + 4*s2)^2)^
(1/2)
 + (4*Z1MA)/(9*a^2 + (s2 + 4*s1)^2)^(1/2)-4/(36*a^2 + (s1 + s2)^2)^
(1/2)
-(ZnA ZnB)/(s1 + s2);
Eq2 = Eb + (3 Z1MA^(1/2))/(s1^2 + a^2)^(3/4);
Eq3 = Eb + (3 Z1MB^(1/2))/(s2^2 + a^2)^(3/4);
Eq4 = Eb + (2*Z1MA)/(s1^2 + a^2)^(1/2) + (2 *Z1MB)/(s2^2 + a^2)
^(1/2)-1/(2 a)
-(ZnA ZnB)/(s1 + s2);
The solution of the system of equations.
Sol = FindRoot[{Eq1 == 0,Eq2 == 0,Eq3 == 0,Eq4 == 0},
{{Eb,3.25},
{s1,.876},{s2,.254},{a,.911}}]
{Eb- > -3.19537,s1- > 0.888324,s2- > 0.261022,a- > 0.922599}
Verification.
N[Eq1]/.Sol
N[Eq2]/.Sol
N[Eq3]/.Sol
N[Eq4]/.Sol
-4.44089*10^-16
-4.44089*10^-16
0.
0.

Calculation of other quantities.
Et/.Sol
r0/.Sol
me/.Sol
-50.4587
1.21642 A
-1.86638 D

B.2.10 The CH Molecule

Output data: E, r_0, μ_e

Input data.
ZA = 6;
ZB = 1;
P = 3.14159265359;
s1A1 = 0.85505;
sA11 = .00269287;
sA1A2 = .25;
alpha = 67;
al = alpha*P/180;
s1A3 = 1/(4*(1/2 + (s1*Cos[al])/(2*(s1^2 + a^2)^(1/2)))^(1/2));
s1A4 = (2)^(1/2)/(4*(1 + (s1*Cos[al]-(3)^(1/2)/2*a*Sin[al])/
(s1^2 + a^2)^(1/2))^(1/2));
s1A5 = (2)^(1/2)/(4*(1 + (s1*Cos[al] + (3)^(1/2)/2*a*Sin[al])/
(s1^2 + a^2)^(1/2))^(1/2));
sA3A4 = 1/(2*(3)^(1/2)*Sin[al]);
Z1A = ZA-2 s1A1-s1A3-s1A4-s1A5;
Z1MA = ZA-2-s1A3-s1A4-s1A5;
Z1MB = 1;
ZA1 = ZA-sA1A2-4*sA11;
ZA3 = ZA-2*s1A1-2*sA3A4-s1A3;
ZA4 = ZA-2*s1A1-2*sA3A4-s1A4;
ZA5 = ZA-2*s1A1-2*sA3A4-s1A5;
Z2B = ZB;
Z2MB = ZB;
excen = .98;
r = (1/ZA3 + 1/ZA4 + 1/ZA5)*2/3*(1 + excen^2/2);
d = r*Cos[al];
b = r*Sin[al];
ZnA = ZA-2-(3*(s1 + s2)^2*(s1 + s2 + d))/((s1 + s2 + d)^2 + b^2)^(3/2);

ZnB = 1;

Et = -(Z1A^2/4)-ZA3^2/4-ZA4^2/4-ZA5^2/4-1-2*ZA1^2
+ 1/((s1/4 + s2/4)^2 + (3/2*a)^2)^(1/2)-1 /((s1/4 + s2)^2 + (3/4 a)^2)^
(1/2)
-(ZA-2)/((s2/4 + s1)^2 + (3/4   a)^2)^(1/2) + 1/((3/4*a)^2 + b^2 +
(s2/4 + s1 + d)^2)^(1/2)
+ 1/((3/4*a-3^(1/2)/2*b)^2 + (b/2)^2 + (s2/4 + s1 + d)^2)^(1/2)
+ 1/((3/4*a + 3^(1/2)/2*b)^2 + (b/2)^2 + (s2/4 + s1 + d)^2)^(1/2)
-3/(b^2 + (s1 + s2 + d)^2)^(1/2) + ((ZA-2)*ZB)/(s1 + s2) + ZA1^
(3/2)/8;

r0 = (s1 + s2)*.529177*2A;

me = -5.08315*(-2*s1 + ZB*(s1 + s2) + 3*d) D;

System of equations.

Eq1 = Eb + Z1A^2/4 + Z2B^2 + (4*Z1MB)/(9*a^2 + (s1 + 4*s2)^2)^
(1/2)
+ (4*Z1MA)/(9*a^2 + (s2 + 4*s1)^2)^(1/2)-4/(36*a^2 + (s1 + s2)^2)^
(1/2)
-(ZnA ZnB)/(s1 + s2);

Eq2 = Eb + (3 Z1MA^(1/2))/(s1^2 + a^2)^(3/4);

Eq3 = Eb + (3 Z1MB^(1/2))/(s2^2 + a^2)^(3/4);

Eq4 = Eb + (2*Z1MA)/(s1^2 + a^2)^(1/2) + (2 *Z1MB)/(s2^2 + a^2)
^(1/2)-1/(2 a)
-(ZnA ZnB)/(s1 + s2);

The solution of the system of equations.

Sol = FindRoot[{Eq1 == 0,Eq2 == 0,Eq3 == 0,Eq4 == 0},
{{Eb,4.3},{s1,.796},{s2,.190},{a,.808}}]

{Eb- > -4.29182,s1- > 0.840475,s2- > 0.175936,a- > 0.767725}

Verification.

N[Eq1]/.Sol

N[Eq2]/.Sol

N[Eq3]/.Sol

N[Eq4]/.Sol

-4.44089*10^-16

0.

0.

4.44089*10^-16

Calculation of other quantities.

Et/.Sol

r0/.Sol

me/.Sol

-76.6349
1.07572 A
-1.91195 D

B.3 PROGRAMS FOR THE CALCULATION OF THE HARMONICS GENERATED AT THE INTERACTION BETWEEN VERY INTENSE LASER FIELDS AND ELECTRONS

We have the following equivalences between the notations from Volume I and this book, and the symbols used in programs: $j \equiv j$, $a_1 \equiv a1$, $a_2 \equiv a2$, $P \equiv number\pi$, $\eta_i \equiv etai$, $\theta \equiv theta$, $\phi \equiv phi$, $\beta_{xi} \equiv betaxi$, $\beta_{yi} \equiv betayi$, $\beta_{zi} \equiv betazi$, $\gamma_i \equiv gammai$, $\eta_1 \equiv eta1$, $\eta_2 \equiv eta2$, $n_x \equiv nx$, $n_y \equiv ny$, $n_z \equiv nz$, $f_0 \equiv f0$, $f_1 \equiv f1$, $f_2 \equiv f2$, $\gamma \equiv gamma$, $f_3 \equiv f3$, $g_1 \equiv g1$, $g_2 \equiv g2$, $g_3 \equiv g3$, $F_1 \equiv F1$, $F_2 \equiv F2$, $h_1 \equiv h1$, $h_2 \equiv h2$, $h_3 \equiv h3$, $\underline{I}_{av} \equiv Iav$, $f_{1sj} \equiv f1sj$, $f_{1cj} \equiv f1cj$, $f_{2sj} \equiv f2sj$, $f_{2cj} \equiv f2cj$, $f_{3sj} \equiv f3sj$, $f_{3cj} \equiv f3cj$, $\underline{I}_j \equiv Ij$, $\gamma_0 \equiv gamma0$, $\beta_0 \equiv beta0$, $\theta' \equiv thetap$, $\phi' \equiv phip$, ω_L(in rad/s) $\equiv omegaL$, $n'_x \equiv nxp$, $n'_y \equiv nyp$, $n'_z \equiv nzp$, $f'_1 \equiv f1p$, $f'_2 \equiv f2p$, $\gamma' \equiv gammap$, $f'_3 \equiv f3p$, $g'_1 \equiv g1p$, $g'_2 \equiv g2p$, $g'_3 \equiv g3p$, $F'_1 \equiv F1p$, $F'_2 \equiv F2p$, $h'_1 \equiv h1p$, $h'_2 \equiv h2p$, $h'_3 \equiv h3p$, $\underline{I}'_j \equiv Ijp$, $f'_{1sjp} \equiv f1sjp$, $f'_{1cjp} \equiv f1cjp$, $f'_{2sjp} \equiv f2sjp$, $f'_{2cjp} \equiv f2cjp$, $f'_{3sjp} \equiv f3sjp$, $f'_{3cjp} \equiv f3cjp$, $\alpha \equiv alpha$, W_j(in keV) $\equiv Wj$, ω_j(in rad/s) $\equiv omegaj$, τ_j(in s) $\equiv tauj$, λ_j(in angstroms) $\equiv lambdaj$, $P \equiv number\ \pi$.

B.3.1 Program for the Calculation of the Harmonics Generated at the Interaction Between Very Intense Laser Beams and Electron Plasmas

Input data: j, a_1, a_2, η_i, θ, ϕ, β_{xi}, β_{yi}, β_{zi}.

Output data: $\underline{I}_{av}$, $\underline{I}_j$.

Input data
j = 53;
a1 = 15;
a2 = 10;
P = 3.14159265358979323846;
etai = P*30/180;
theta = P*4.182/180;
phi = P*-15.1/180;
betaxi = 0.15;
betayi = -0.10;
betazi = 0.20;

gammai = 1/(1-betaxi^2-betayi^2-betazi^2)^(1/2);
Calculation of average intensity
eta1 = 0;
eta2 = 2*P;
nx = Sin[theta]*Cos[phi];
ny = Sin[theta]*Sin[phi];
nz = Cos[theta];
f0 = gammai*(1-betazi);
f1 = -a1*(Sin[eta]-Sin[etai]) + gammai*betaxi;
f2 = -a2*(Cos[etai]-Cos[eta]) + gammai*betayi;
gamma = 1/(2*f0)*(1 + f0^2 + f1^2 + f2^2);
f3 = gamma-f0;
g1 = -((a1*f0)/gamma^2)*Cos[eta]
  + f1/gamma^3*(a1*f1*Cos[eta] + a2*f2*Sin[eta]);
g2 = -((a2*f0)/gamma^2)*Sin[eta]
  + f2/gamma^3*(a1*f1*Cos[eta] + a2*f2*Sin[eta]);
g3 = -(f0/gamma^3)*(a1*f1*Cos[eta] + a2*f2*Sin[eta]);
F1 = 1-(nx*f1)/gamma-(ny*f2)/gamma-(nz*f3)/gamma;
F2 = nx*g1 + ny*g2 + nz*g3;
h1 = -F1*g1 + F2*(nx-f1/gamma);
h2 = -F1*g2 + F2*(ny-f2/gamma);
h3 = -F1*g3 + F2*(nz-f3/gamma);
F = 1/F1^6*(h1^2 + h2^2 + h3^2);
Iav = NIntegrate[F/(2*Pi),{eta,eta1,eta2}]
1.11223*10^11
Calculation of spectral intensity
f1sj = NIntegrate[(h1/(P*F1^3))*Sin[j*eta],{eta,eta1,eta2},
AccuracyGoal- > 12];
f1cj = NIntegrate[(h1/(P*F1^3))*Cos[j*eta],{eta,eta1,eta2},
AccuracyGoal- > 12];
f2sj = NIntegrate[(h2/(P*F1^3))*Sin[j*eta],{eta,eta1,eta2},
AccuracyGoal- > 12];
f2cj = NIntegrate[(h2/(P*F1^3))*Cos[j*eta],{eta,eta1,eta2},
AccuracyGoal- > 12];
f3sj = NIntegrate[(h3/(P*F1^3))*Sin[j*eta],{eta,eta1,eta2},
AccuracyGoal- > 12];
f3cj = NIntegrate[(h3/(P*F1^3))*Cos[j*eta],{eta,eta1,eta2},
AccuracyGoal- > 12];
Ij = 1/2*(f1sj^2 + f1cj^2 + f2sj^2 + f2cj^2 + f3sj^2 + f3cj^2)
1.30467*10^8

B.3.2 Program for the Calculation of the Harmonics Generated at the Head-On Interaction Between Very Intense Laser Beams and Relativistic Electron Beams

Input data: j, a_1, a_2, η_i, θ', ϕ', ω_L, θ_L.

Output data: $\underline{I}'_j$, $\underline{I}_j$, θ, α, W_j, ω_j, τ_j and λ_j.

Input data.
al = 7.5;
a2 = 0;
j = 19;
gamma0 = 195.8;
beta0 = (1-1/gamma0^2)^(1/2);
P = 3.14159265358979323846;
etai = P*0/180;
thetap = P*180/180;
phip = P*0/180;
omegaL = 2.355*10^15;
Calculation of the intensity of the scattered beam in the S' system
eta1 = 0;
eta2 = 2*P;
nxp = Sin[thetap]*Cos[phip];
nyp = Sin[thetap]*Sin[phip];
nzp = Cos[thetap];
f1p = -a1*(Sin[eta]-Sin[etai]);
f2p = -a2*(Cos[etai]-Cos[eta]);
gammap = 1/2*(2 + f1p^2 + f2p^2);
f3p = gammap-1;
g1p = -(a1/gammap^2)*Cos[eta] + f1p/gammap^3*(a1*f1p*Cos
[eta] + a2*f2p*Sin[eta]);
g2p = -(a2/gammap^2)*Sin[eta] + f2p/gammap^3*(a1*f1p*Cos
[eta] + a2*f2p*Sin[eta]);
g3p = -((a1*f1p*Cos[eta] + a2*f2p*Sin[eta])/gammap^3);
F1p = 1-(nxp*f1p)/gammap-(nyp*f2p)/gammap-(nzp*f3p)/gammap;
F2p = nxp*g1p + nyp*g2p + nzp*g3p;
h1p = -F1p*g1p + F2p*(nxp-f1p/gammap);
h2p = -F1p*g2p + F2p*(nyp-f2p/gammap);
h3p = -F1p*g3p + F2p*(nzp-f3p/gammap);
f1sjp = NIntegrate[(h1p/F1p^3)*Sin[eta*j],{eta,eta1,eta2},
AccuracyGoal- > 12];

f1cjp = NIntegrate[(h1p/F1p^3)*Cos[eta*j],{eta,eta1,eta2},
AccuracyGoal- > 12];
f2sjp = NIntegrate[(h2p/F1p^3)*Sin[eta*j],{eta,eta1,eta2},
AccuracyGoal- > 12];
f2cjp = NIntegrate[(h2p/F1p^3)*Cos[eta*j],{eta,eta1,eta2},
AccuracyGoal- > 12];
f3sjp = NIntegrate[(h3p/F1p^3)*Sin[eta*j],{eta,eta1,eta2},
AccuracyGoal- > 12];
f3cjp = NIntegrate[(h3p/F1p^3)*Cos[eta*j],{eta,eta1,eta2},
AccuracyGoal- > 12];
Ijp = 1/2*(f1sjp^2 + f1cjp^2 + f2sjp^2 + f2cjp^2 + f3sjp^2 + f3cjp^2)
0.767474
Calculation of the intensity of the scattered beam in the S system
Ij = gamma0^2*(1-beta0*Cos[thetap])^2*Ijp
117691.
Calculation of theta (in rad), alpha (in mrad), Wj (in keV), omegaj
(in rad/s), tauj (in s) and lambdaj (in angstroms)
theta = N[ArcCos[(Cos[thetap]-beta0)/(1-Cos[thetap]*beta0)]]
alpha = N[1000*(P-ArcCos[(Cos[thetap]-beta0)/(1-Cos[thetap]*
beta0)])]
Wj = j*1.0545716*10^-34*omegaL*gamma0^2*(1 + beta0)*(1-
beta0*Cos[thetap])/(1000*1.602176*10^-19)
omegaj = j*omegaL*gamma0^2*(1 + beta0)*(1-beta0*Cos[thetap])
tauj = (2*P)/omegaj
lambdaj = (2*P*2.998*10^8)/(j*omegaL*gamma0^2*(1 + beta0)*
(1-beta0*Cos[thetap]))*10^10
3.14159
0.
4516.38
6.86158*10^21
9.15705*10^-22
0.00274528

B.4 PROGRAM FOR THE CALCULATION OF THE HARMONICS GENERATED AT THE INTERACTION BETWEEN LASER BEAM AND ATOMS

We have the following equivalences between the notations from Volume I and this book, and the symbols used in programs: $\eta_0 \equiv eta0$,

$\eta_1 \equiv eta1$, $I_p(\text{in eV}) \equiv Ip$, $I_L(\text{in } W/m^2) \equiv IL$, $\lambda_L(\text{in m}) \equiv lambdaL$, $m_m \equiv mm$, $c(\text{in } m/s) \equiv c$, $\varepsilon_0(\text{in } F/m) \equiv eps0$, $m(\text{in kg}) \equiv m$, $e(\text{in C}) \equiv q$, $I_H(\text{in eV}) \equiv IH$, $a_0(\text{in m}) \equiv a0$, $\hbar(\text{in } m^2 \text{ kg s}) \equiv hbar$, $\omega_L(\text{in } rad/s) \equiv omegaL$, $E_M(\text{in } V/m) \equiv EM$, $a \equiv a$, $U_p(\text{in eV}) \equiv Ep$, $E_q(\text{in eV}) \equiv Eq$, $E_c(\text{in eV}) \equiv Ec$, $n_c \equiv nc$, $\gamma_1 \equiv gamma1$, $n^* \equiv ns$, $\underline{W} \equiv Wn$, $K \equiv K$, $f \equiv f$, $\underline{R} \equiv Rn$, $n \equiv n$, $E_{em}(\text{in eV}) \equiv Eem$, $\underline{\eta}_1 \equiv eta1n$, $\underline{E}_k \equiv Ekn$, $P \equiv number\ \pi$.

Input data: η_0, η_1, I_p, I_L, λ_L.

Output data: $\underline{r}$, n, E_{em}, $\underline{\eta}_1$, $\underline{E}_k$.

Calculation of eta1.
P = 3.14159265358979323846;
eta0 = N[(1.3*P)/180];
Eq1 = Cos[x]-Cos[eta0] + Sin[eta0]*(x-eta0);
Sol = FindRoot[{Eq1 = = 0},{{x,5}}]
N[Eq1]/.Sol;
{x- > 5.76644}
Input data.
eta0 = N[(0.18*P)/180];
eta1 = 6.08733;
Ip = 15.7596;
IL = 3*10^21;
lambdaL = .248*10^-6;
mm = 0;
Constants and parameters.
c = 2.9979*10^8;
eps0 = 8.85419*10^-12;
m = 9.10938*10^-31;
q = 1.60218*10^-19;
IH = 13.60569;
a0 = 5.29177*10^-11;
hbar = 1.05457*10^-34;
omegaL = (2*P*c)/lambdaL;
EM = ((2*IL)/(eps0*c))^(1/2);
a = (q*EM)/(m*c*omegaL);
Ep = (q^2*EM^2)/(4*m*omegaL^2*1.6021*10^-19);
Eq = omegaL*hbar/(1.6021*10^-19);
Ec = 3.17*Ep + Ip;
nc = Ec/(omegaL*hbar/(1.6021*10^-19));

gamma1 = (2*Ep)/Ip;

Calculation of rate of ionization

ns = (IH/Ip)^(1/2);

Wn = 1/(Cos[eta0])^(2*ns-mm-1)*Exp[-(1/Cos[eta0]-1) (4*(Ip*1.6021*
10^-19)^(3/2))/(3*q*EM*(2*m)^(-1/2)*hbar)]

1.

Calculation of overall rate of generation of harmonics

K = 2^7/P^2*(q^2/(4*P*eps0*a0))^(5/2);

f = gamma1*(Sin[eta1]-Sin[eta0])^2*(1 + a^2/4   (Sin[eta1]-Sin[eta0])
^2);

Rn = K*Wn*(2*Ip*(1.6021*10^-19))^(-(5/2)) f/(f + 1)^6

n = (0.5*a^2*m*c^2*(Sin[eta1]-Sin[eta0])^2 + Ip*1.60218*10^-19)/
(hbar*omegaL)

Eem = (0.5*a^2*m*c^2*(Sin[eta1]-Sin[eta0])^2)/(1.60218*10^-19) +
Ip

0.000101164

30.1048

150.503

Calculation of eta1n and Ekn.

eta1n = eta1/(2*P);

Ekn = (2/3.17)*(Sin[eta1]-Sin[eta0])^2;

BIBLIOGRAPHY

Agostini, P., DiMauro, L.F., 2008. Contemp. Phys. 49, 179.

Anderson, S.G., Barty, C.P.J., Betts, S.M., et al., 2004. Appl. Phys. B, 78, 891.

Arfken, G., 1985. Mathematical Methods for Physicists, third ed. Academic Press, Orlando, FL.

Babzien, M.E., Ben-Zvi, I., Kusche, K., 2006. Phys. Rev. Lett. 96, 054802.

Beebe, N.H.F., Lunell, S., 1975. J. Phys. B At. Mol. Opt. Phys. 8, 2320.

CCCBDB. The Computational Chemistry Comparison and Benchmark Database. <http://cccbdb.nist.gov/>.

Clementi, E., Roetti, C., 1974. At. Data Nucl. Data Tab. 14, 177.

Corkum, P.B., 1993. Phys. Rev. Lett. 71, 1994.

Coulson, C.A., 1961. Valence. Oxford University Press, London.

de Castro, E.V.R., Jorge, F.E., 2001. An. Acad. Bras. Cienc. 73, 511.

Ferray, M., L'Huillier, A., Li, X.F., Lompre, L.A., Mainfray, G., Manus, C., 1988. J. Phys. B: At. Mol. Opt. Phys. 21, L31.

Goreslavskii, S.P., Popruzhenko, S.V., Shcherbachev, O.V., 1999. Laser Phys. 9, 1039.

Gryzinski, M., 1973. Phys. Lett. 44A, 131.

Hartree, D.R., 1957. The Calculation of Atomic Structures. John Wiley, New York, NY.

Hertzberg, G., 1950. Molecular Spectra and Molecular Structure. I. Spectra of Diatomic Molecule. Van Nostrand, New York, NY.

Huber, K.P., Hertzberg, G., 1979. Molecular Spectra and Molecular Structure. IV. Constants of Diatomic Molecules. Van Nostrand, New York, NY.

Huzinaga, S., Arnau, C., 1970. J. Chem. Phys. 53, 451.

Kim, K.J., Chattopadhyay, S., Shank, C.V., 1994. Nucl. Instrum. Methods Phys. Res. A 341, 351.

Koga, T., Tatewaki, H., Thakkar, A.J., 1993. Phys. Rev. A 47, 4510–4512.

Krause, J.L., Schafer, K.J., Kulander, K.C., 1992. Phys. Rev. Lett. 68, 3535.

Ladner, R.C., Goddard III, W.A., 1969. J. Chem. Phys. 51, 1073.

Landau, L.D., Lifschitz, E.M., 1991. Quantum Mechanics. Pergamon Press, New York, NY.

Landau, L.D., Lifschitz, E.M., 2000. Mechanics. Butterworth Heinemann, Oxford.

Lewenstein, M., Balcou, P., Ivanov, M.Y., et al., 1994. Phys. Rev. A 49, 2117.

Lide, D.R., 2003. CRC Handbook of Chemistry and Physics. CRC Press, Boca Raton, FL.

Pauling, L., 1970. General Chemistry. W. H. Freeman and Company, San Francisco, CA.

Pogorelsky, I.V., Ben-Zvi, I., Hirose, T., et al., 2000. Phys. Rev. S. T. Accel. Beams 3, 090702.

Popa, A., 1996. Rev. Roum. Mathem. Pures Appl. 41, 109.

Popa, A., Lazarescu, M., Dabu, R., Stratan, A., 1997. IEEE J. Quantum Electron. 33, 1474.

Popa, A., 1998a. Rev. Roum. Mathem. Pures Appl. 43, 415–424.

Popa, A., 1998b. J. Phys. Soc. Jpn. 67, 2645.

Popa, A., 1999a. Rev. Roum. Math. Pures Appl. 44, 119.

Popa, A., 1999b. J. Phys. Soc. Jpn. 68, 763.

Popa, A., 1999c. J. Phys. Soc. Jpn. 68, 2923.

Popa, A., 2000. In: Kajzar, F., Agranovich, M.V. (Eds.), Multiphoton and Light Driven Multielectron Processes in Organics: New Phenomena, Materials and Applications. Kluwer Academic Publishers, Amsterdam, p. 513.

Popa, A., 2003a. J. Phys. A Math. Gen. 36, 7569.

Popa, A., 2003b. J. Phys. Condens. Matter. 15, L559.

Popa, A., 2004. IEEE J. Quantum Electron. 40, 1519.

Popa, A., 2005. J. Chem. Phys. 122, 244701.

Popa, A., 2007. IEEE J. Quantum Electron. 43, 1183.

Popa, A., 2008a. Eur. Phys. J. D 49, 279.

Popa, A., 2008b. J. Phys. B At. Mol. Opt. Phys. 41, 015601.

Popa, A., 2009a. Eur. Phys. J. D 54, 575.

Popa, A., 2009b. J. Phys. B At. Mol. Opt. Phys. 42, 025601.

Popa, A., 2011a. Mol. Phys. 109, 575.

Popa, A., 2011b. Phys. Rev. A 84, 023824.

Popa, A., 2011c. Proc. Romanian Acad. A 12, 302.

Popa, A., 2012. Laser Part. Beams 30, 591.

Poulsen, M.D., Madsen, L.B., 2005. Phys. Rev. A 72, 042501.

Sarukura, N., Hata, K., Adachi, T., Nodomi, R., Watanabe, M., Watanabe, S., 1991. Phys. Rev. A, 43, 1669.

Slater, J.C., 1960. *Quantum Theory of Atomic Structure, vols. 1 and 2*. McGraw-Hill, New York, NY.

Slater, J.C., 1963. Quantum Theory of Molecules and Solids, vol. 1. McGraw-Hill, New York, NY.

Printed and bound by CPI Group (UK) Ltd, Croydon, CR0 4YY

03/10/2024

01040423-0014